AF735504

The Foundations
of
Einstein's Theory of Gravitation

The Foundations
of
Einstein's Theory of Gravitation

by

ERWIN FREUNDLICH

Authorised English Translation by

HENRY L. BROSE, M.A.
Christ Church, Oxford

Preface by

ALBERT EINSTEIN

Introduction by

H. H. TURNER, D.Sc., F.R.S.
Savilian Professor of Astronomy in the University of Oxford

CAMBRIDGE
AT THE UNIVERSITY PRESS
1920

CAMBRIDGE UNIVERSITY PRESS
Cambridge, New York, Melbourne, Madrid, Cape Town,
Singapore, São Paulo, Delhi, Tokyo, Mexico City

Cambridge University Press
The Edinburgh Building, Cambridge CB2 8RU, UK

Published in the United States of America by Cambridge University Press, New York

www.cambridge.org
Information on this title: www.cambridge.org/9781107601376

First published 1920
Re-issued 2011

A catalogue record for this publication is available from the British Library

ISBN 978-1-107-60137-6 Paperback

PREFACE

DR FREUNDLICH has undertaken in the following essay to illumine the ideas and observations which gave rise to the general theory of relativity so as to make them available to a wider circle of readers.

I have gained the impression in perusing these pages that the author has succeeded in rendering the fundamental ideas of the theory accessible to all who are to some extent conversant with the methods of reasoning of the exact sciences. The relations of the problem to mathematics, to the theory of knowledge, physics and astronomy are expounded in a fascinating style, and the depth of thought of Riemann, a mathematician so far in advance of his time, has in particular received warm appreciation.

Dr Freundlich is not only highly qualified as a specialist in the various branches of knowledge involved to demonstrate the subject; he is also the first amongst fellow-scientists who has taken pains to put the theory to the test.

May his booklet prove a source of pleasure to many!

A. EINSTEIN.

TRANSLATOR'S NOTE

FOR the sake of those English readers who wish to pursue the development of the special (or restricted) and the general theory of relativity in greater detail, reference may be made to the following:

"On the Hypotheses which lie at the Bases of Geometry," being Bernh. Riemann's disquisition, translated by W. K. Clifford in *Nature* (1873), vol. VIII. Nos. 183, 184, pp. 14–17, 36, 37.

Or in the *Mathematical Papers* of W. K. Clifford, pp. 55–71.

(The Introduction to the latter volume and the paper "On the Classification of Loci" bear upon the question.)

"Memoir on Abstract Geometry." Cayley, *Math. Papers*, vol. VI. 1893, No. 413.

The special theory of relativity is dealt with in:

The Principle of Relativity. Eb. Cunningham. Camb. Univ. Press.

Relativity and the Electron Theory. Eb. Cunningham. Monographs on Physics. Longmans, Green & Co.

The Principle of Relativity. L. Silberstein. Macmillan & Co.

"The Space-Time Manifold of Relativity." E. B. Wilson and G. N. Lewis. *Proc. of American Assoc. of Arts and Science*, vol. XLVIII. No. 11, 1912.

The general theory of relativity has been excellently summarised by Prof. A. S. Eddington in his "Report on the Relativity Theory of Gravitation for the Physical Society of London," published by the Fleetway Press, Ltd., Fleet St.

Vide also Prof. W. de Sitter, "On Einstein's Theory of Gravitation and its Astronomical Consequences" in the *Monthly Notices of the Royal Astronomical Society* in three papers:

Vol. LXXVI. p. 699. Vol. LXXVII. p. 155. Vol. LXXVIII. Nov. 1917.

Prof. J. H. Jeans, "Recent Developments of Molecular Physics" in the *Report of Royal Inst. of Great Britain*, Friday, March 30, 1917.

Prof. A. S. Eddington. Article in *Nature*, vol. XCVIII. p. 328 (Dec. 28, 1916).

Articles by Dr H. Wildon Carr, Prof. F. A. Lindemann, and Prof. A. N. Whitehead in the Educational Supplement of *The Times* (Jan. 22 and 29, Feb. 5, 1920).

The monthly copies of *The Observatory* (published by Taylor and Francis, Red Lion Court, Fleet St.) contain a number of interesting reviews and discussions about the theory of gravitation of which the following may be specially mentioned:

The Observatory, No. 505, p. 412. "Space, Time and Gravitation," by Prof. W. de Sitter.

The Observatory, March, 1919. "The Total Eclipse of 29 May, 1919, and the Influence of Gravitation on Light," by Prof. A. S. Eddington.

The following quotation may assist in making intelligible one of the vital points which tends to mystify the unphilosophical physicist:

"When a rod is started from rest into uniform motion, nothing whatever happens to the rod. We say that it contracts; but length is not a property of the rod; it is a relation between the rod and the observer. *Until the observer is specified the length of the rod is quite indeterminate.* We ought always to remember that our experiments reveal only relations, and not properties inherent in individual objects; and then the correspondence of two systems, differing only in uniform motion, becomes axiomatic, so that laborious mathematical verifications are redundant." (Prof. Eddington.)

The introductory words of Minkowski's famous paper may be recalled in conclusion:

"From henceforth time by itself and space by itself are mere shadows, they are only two aspects of a single and indivisible manner of co-ordinating the facts of the physical world." (Minkowski.)

I wish to express my thanks to Consul Arnold Gumprecht and Miss Gertrud Gerdau for valuable assistance in the earlier stages of translation, to Dr Freundlich for his kindness in perusing the manuscript, and to Professors Turner and Eddington for the great interest they have shewn in getting the booklet published. I am indebted to Mr J. W. N. Smith, M.A., for kindly reading the proofs.

HENRY L. BROSE.

CHRIST CHURCH,
OXFORD.
2 *February*, 1920.

CONTENTS

INTRODUCTION

THE Universe is limited by the properties of light. Until half a century ago it was strictly true that we depended upon our eyes for all our knowledge of the universe, which extended no further than we could see. Even the invention of the telescope did not disturb this proposition, but it is otherwise with the invention of the photographic plate. It is now conceivable that a blind man, by taking photographs and rendering their records in some way decipherable by his fingers, could investigate the universe; but still it would remain true, that all his knowledge of anything outside the earth would be derived by the use of light and would therefore be limited by its properties. On this little earth there is indeed a tiny corner of the universe accessible to other senses: but feeling and taste act only at those minute distances which separate particles of matter when "in contact": smell ranges over, at the utmost, a mile or two; and the greatest distance which sound is ever known to have travelled (when Krakatoa exploded in 1883) is but a few thousand miles—a mere fraction of the earth's girdle. The scale of phenomena manifested through agencies other than light is so small that we are unlikely to reach any noteworthy precision by their study.

Few people who are not astronomers have spent much thought on the limitations introduced by the news agency to which we are so profoundly indebted. Light comes speedily but has far to travel, and some of the news is thousands of years old before we get it. Hence our universe is not co-existent: the part close around us belongs to the peaceful present, but the nearest star is still in the midst of the late War, for our news of him is three years old; other stars are Elizabethan, others belong to the time of the Pharaohs; and we have alongside our modern civilization yet others of prehistoric date. The electric telegraph has accustomed us to a world in which the news is approximately of even date: but our forefathers must have been better able, from their daily experience of getting news many months old, to realise the unequal age of the universe we know. Nowadays the inequality is almost entirely the concern of the astronomer, and even he often neglects or forgets it. But when fundamental issues are at stake, the time taken by the messenger is an essential part of the discussion, and we must be careful to take account of it, with the utmost precision.

Our knowledge that light had a finite velocity followed on the invention of the telescope and the discovery of Jupiter's satellites: the news of their eclipses came late at times and these times were identified as those when Jupiter was unusually far away from us. But the full consequences of the discovery were not realised at first. One such consequence is that the stars are not seen in their true places, that is in the places which they truly held when the light left them (for what may have happened to them since we do not know at all—they may have gone out or exploded). Our earth is only moving slowly compared with the great haste of light: but still she is moving, and consequently there is "aberration"—a displacement due to the ratio of the two velocities, easy enough to recognise now, but so difficult to apprehend for the first time that Bradley spent two years in worrying over the conundrum presented by his observations before he thought of the solution. It came to him unexpectedly, as often happens in such cases. In his own words—"at last when he despaired of being able to account for the phenomena which he had observed, a satisfactory explanation of them occurred to him all at once when he was not in search of it." He accompanied a pleasure party in a sail upon the river Thames. The boat in which they were was provided with a mast which had a vane at the top of it. It blew a moderate wind, and the party sailed up and down the river for a considerable time. Dr Bradley remarked that every time the boat put about, the vane at the top of the boat's mast shifted a little, as if there had been a slight change in the direction of the wind. The sailors told him that this was due to the change in the boat, not the wind: and at once the solution of his problem was suggested. The earth running hither and thither round the sun resembles the boat sailing up and down the river: and the apparent changes of wind correspond to the apparent changes in direction of the light of a star. But now comes a point of detail—does the vane itself affect the wind just round it? And, similarly, does the earth itself by its movement affect the ether just round it, or the apparent direction of the light waves? This question suggested the famous Michelson and Morley experiment (*Phil. Mag.* Dec. 1887). It is curious to think that in the little corner of the universe represented by the space available in a laboratory an experiment should be possible which alters our whole conceptions of what happens in the profoundest depths of space known to us, but so it is. The laboratory experiment of Michelson and Morley was the first step in the great advance recently made. It discredited the existence of the virtual stream of ether which is the natural antithesis to the earth's actual motion. It was indeed open to question whether restrictions of a laboratory might not be responsible for the result: for the ether stream might exist, but the laboratory in which it was hoped to detect it might be in a sheltered eddy. When bodies move through the air, they encounter an apparent stream of opposing air, as all motorists know: but by using a glass screen shelter from the stream can be found. And even without such special screening, there may be shelter. When a pendulum is set swinging in ordinary air, it is found from experiments on clocks that it carries a certain amount of air along with it in its movement, although the portion carried probably clings closely to the surface of the pendulum. A very small insect placed in the region might be unable to detect the streaming of the air further out. In a similar way it seemed possible that as the earth moved through the ether such tiny insects as the physicists in their laboratories might be in a part of the ether carried along with the earth, in which they could not detect the streaming outside. But another laboratory experiment, this time by Sir Oliver Lodge, discredited this explanation, and it was then suggested as an alternative that distances were automatically altered by movement.

It may be well to explain briefly the significance of this alternative. The Michelson-Morley experiment depended on the difference between travelling up-and-down-stream, and across it. To use a few figures may be the quickest way of making the point clear. Suppose a very wide, perfectly smooth stream running at three miles an hour, and that oarsmen are to start from a *fixed* point *O* in mid-stream, row out in any direction to a distance of four miles from *O*, and back again to the starting-point *O*. Which is the best direction to choose? We shall probably all agree that it will be either directly up and down stream, or directly across it, and we may confine attention to these two directions. First suppose an oarsman *A* starts straight across stream. To keep straight he must set his boat at an angle to the stream. If he reaches his four mile limit in an hour, the stream has been virtually carrying him down three miles in a direction at right angles to his course: and the well-known relation between the sides of a right-angled triangle tells us that he has effectively pulled five miles in the hour. It will take him similarly an hour to come back, and the total journey will involve an effective pull of ten miles.

Now suppose another oarsman *B* of equal skill elects to row up stream. In two hours he could pull ten miles if there were no stream; but since meantime the stream has pulled him back six miles by "direct action" he will have only just reached the four mile limit from the start, and has still his return journey to go. No doubt he will accomplish this pretty quickly with the stream to help him, but his antagonist has already got home before he begins the return. We might have let him do his quick journey down stream first, but it is easy to see that this would gain him no ultimate advantage.

Michelson and Morley sent two rays of light on two journeys similar to those of the oarsmen *A* and *B*. The stream was the supposed stream of ether from east to west which should result from the earth's movement of rotation from west to east. They confidently expected the return of *A* before that of *B*, and were quite taken aback to find the two reaching the goal together. In the aquatic analogy of which we have made use, it would no doubt be suspected that *B* was really the faster oar, which might be tested by interchanging the courses; but there are no known differences in the velocity of light which would allow of a parallel explanation. There was however the possibility that the distances had been marked wrongly, and this was tested by interchanging them, without altering the "dead-heat."

Now there are several alternative explanations of this result. One is that the ether does not itself exist, and therefore there is no stream of it, actual or apparent; and it is to this sweeping conclusion that modern reasoning, following recent experiments and observations, is tending. The possibility of saving the ether by endowing it with four dimensions instead of three is scarcely calculated to satisfy those who believed (until recently) that we knew more about the ether than about matter itself. They saved the ether for a time by an automatic shortening of all bodies in the direction of their movement, which explained the dead-heat puzzle. With the velocities used above the goal attained by *B* must be automatically moved $\frac{4}{5}$ of a mile nearer the starting-point, so that *B* only rows $3\frac{1}{5}$ miles out and back instead of four miles. So gross a piece of cheating would enable *B* to make his dead-heat, but could scarcely escape detection. The shortening of the course required in the case of light is very minute indeed, because the velocities of the heavenly bodies are so small compared with that of light. If they could be multiplied a thousand times we might see some curious things but we have no actual experience to guide a forecast.

It is a great triumph for Pure Mathematics that it should have devised a forecast for us in its own peculiar way. Starting from axioms or postulates, Einstein by sheer mathematical skill, making full use of the beautiful theoretical apparatus inherited from his predecessors, pointed ultimately to three observational tests, three things which must happen if the axioms and postulates were well founded. One of the tests—the movement of the perihelion of Mercury's orbit—had already been made and was awaiting explanation as a standing puzzle. Another—a displacement of lines in the spectrum of the sun—is still being made, the issue being not yet clear.

The third suggestion was that the rays of light from a star would be bent on passing near the sun by a particular amount, and this test has just provided a sensational triumph for Einstein. The application was particularly interesting because it was not known which of at least three results might be attained. If light were composed of material particles as Newton suggested, then in passing the sun they would suffer a natural deflection (the use of the adjective is an almost automatic consequence of modes of thought which we must now abandon) which we may call N. On Einstein's theory the deflection would be just twice this amount, $E = 2N$. But it was thought quite possible that the result might be neither N nor E but zero, and Professor Eddington remarked before setting out on the recent expedition that a zero result, however disappointing immediately, might ultimately turn out the most fruitful of all. That was less than a year ago. Perhaps a few dates are worth remembering. Einstein's theory was fully developed and stated in November, 1915, but news of it did not reach England (owing to the War) for some months. In 1917 the Astronomer Royal pointed out the special suitability of the Total Solar Eclipse of May, 1919, as an occasion for testing Einstein's Theory. Preparations for two Expeditions were commenced—Mr Hinks described the geographical conditions on the central line in November, 1917—but could not be fully in earnest until the Armistice of November, 1918. In November, 1919, the entirely satisfactory outcome was announced to the Royal Society and characterised by the President as necessitating a veritable revolution in scientific thought.

But when Mr Brose brought me his translation of the pamphlet in the spring of 1919, the issue was still in doubt. He had become deeply interested in the new theory while interned in Germany as a civilian prisoner and had there made this translation. I encouraged him to publish it and opened negotiations to that end, but it was not until we enlisted the sympathy of Professor Eddington (on his return from the Expedition) and approached the Cambridge Press that a feasible plan of publication was found. Professor Eddington would have been a far more appropriate introducer; and it is only in deference to his own express wish that I have ventured to take up the pen that he would have used to much better purpose. One advantage I reap from the decision: I can express the thanks of Mr Brose and myself to him for his practical help, and perhaps I may add those of a far wider circle for his own able expositions of an intricate theory, which have done so much to make it known in England.

H. H. TURNER.

UNIVERSITY OBSERVATORY,
OXFORD.
November 30th, 1919.

§ 1

THE *SPECIAL* THEORY OF RELATIVITY AS A STEPPING-STONE TO THE *GENERAL* THEORY OF RELATIVITY

In the following pages the results of the special principle of relativity will often be utilised. In order not to have to interrupt the course of discussion later, I shall therefore open with a chapter dealing with the significance of the *special** principle of relativity as a stepping-stone to the *general* principle. The difficulties involved in the principles of classical mechanics are to be treated separately in a later chapter: they will therefore only be considered here as far as is absolutely necessary.

The entire upheaval, which we are witnessing in the world of physics at the present time, received its impulse from obstacles which were encountered in the progress of electrodynamics. Starting from these, therefore, we shall best be able to trace how a series of new discoveries, brought to light by electrodynamics, had necessarily to lead to an entirely new view of the foundations of mechanics. Most of the objections against the latter developments, it is true, have been raised for the very reason that a branch of science, which was not considered to have a just claim to deal with questions of mechanics, asserted the right of exercising a far-reaching influence upon the latter extending to its foundations even.

If, however, we trace these objections to their source, we discover that they are due to a wish to give mechanics the form of a purely *mathematical science*, similar to geometry, in spite of the fact that it is founded upon hypotheses which are essentially *physical*: up to the present, certainly, these hypotheses have not been recognised to be such.

The development of electrodynamics took place essentially without being influenced by the results of mechanics, and without exerting any influence itself upon the latter, so long as its range of investigation remained confined to the electrodynamical phenomena

* *Vide* Note 1 (p. 43).

of bodies *at rest*. Only after Maxwell's equations had supplied a foundation for these, did it become possible to take up the study of the electrodynamical phenomena of *moving* media. Numerous questions of far-reaching importance hereupon intruded themselves. All optical occurrences—and according to Maxwell's theory these also belong to the sphere of electrodynamics—take place either between bodies, which are in motion relatively to one another as in astrophysics, or upon the earth, which revolves about the sun with a velocity of about 30 kilometres per second, and performs, together with the sun, a translational motion of about the same order of magnitude. Accordingly, it was a problem of great importance to ascertain the influence of the motion of a source of light upon the velocity of propagation of the light it emits. An endeavour was therefore made to find a theory of these phenomena in which electrodynamical and mechanical effects occurred simultaneously. Mechanics, which had long stood as a structure complete in every detail, had to stand the test whether it was capable of supplying the fitting arguments for a description of these *complex* phenomena.

In addition to these optical problems, however, the discovery of electrons revealed new facts, in which electrodynamical and mechanical phenomena take place together. In the case of cathode-rays and radium preparations we observe the motions of elementary particles of electricity, *free from the action of mass-attraction* (gravitation). We have here to a certain extent a counterpart to the fundamental observations upon which mechanics is founded, viz. to the motions of the heavenly bodies: these take place uninfluenced by any other physical actions, solely under the influence of mass-attraction. From this quarter, too, mechanics was subjected to a crucial test (proof-test).

The first outstanding attempt to describe these phenomena was made by H. Hertz. He extended Maxwell's equations by additional terms, which were to represent the influence of the motion of electric charges. Hertz made no definite assumptions about the character of the motion of electric charges in his extensions, and was fully aware that his new equations would not encompass the *optical* phenomena produced by moving media. Lorentz's electrodynamics first led to fundamental equations which agreed with the results of observation and experiment. From these fundamental equations again the problem of finding a relativity-principle, corresponding to the equations describing physical phenomena, evolved itself. Up to this point the Galilei-Newton relativity-principle of classical mechanics had remained master of the field. This principle requires that two systems of coordinates, moving with uniform motion in a straight line with respect to one another, are to be regarded as fully equivalent for the description of events in the domain of mechanics. The transition from one such system to another is effected by means of equations of transformation of the form:

$$x' = x - vt,\ y' = y,\ z' = z,\ t' = t \quad \ldots\ldots\ldots\ldots(1),$$

if we restrict ourselves, for the sake of simplicity, to the case of two systems moving relatively to one another with a velocity v along their x-axes.

The study of the electrodynamical phenomena of moving bodies, however, finally led to a system of fundamental equations, which did not remain invariable after the application of transformations of the *above* form, but only after transformations of the form:

$$x' = \frac{x - vt}{\sqrt{1 - \frac{v^2}{c^2}}},\ y' = y,\ z' = z,\ t' = \frac{t - \frac{v}{c^2} \cdot x}{\sqrt{1 - \frac{v^2}{c^2}}} \quad \ldots\ldots(2),$$

(where c denotes the velocity of light *in vacuo*). These are the so-called Lorentz-Einstein transformations. From this one could not but conclude that the relativity-principle of Newton and Galilei does not hold for electrodynamics. Does it entail entirely giving up the relativity-principle in physics? As a result of repeated fruitless attempts to demonstrate the motion of the earth by means of optical experiments, the impression grew increasingly stronger that there must inevitably be a principle of relativity in physical nature, and that absolute motion has no meaning in itself. But as there can be only *one* principle of relativity in the equations both of physics and of mechanics, the relativity-principle of mechanics had to adapt itself to that which had been derived from the equations of electrodynamics. This constitutes the so-called Lorentz-Einstein *special* theory of relativity, which also—like the relativity-principle of Newton and Galilei—postulates the *equiva-*

lence of two systems of coordinates, which are in uniform rectilinear motion with respect to one another, in order to describe physical events mathematically: but connects the coordinates x, y, z, t and x', y', z', t' of two such systems with one another by means of the Lorentz-Einstein transformation-formulae (2) given above.

From these remarks we see that the two principles of relativity (i.e. the mechanical and the electrodynamical) differ from one another only in the form of the group of mathematical transformation-formulae, by which such equivalent systems are mutually related.

At this stage the question forces itself upon one: what physical assumptions are involved in the transformations of the relativity-principle of Newton and Galilei, and in what respect do they so far disagree with experience or experiment that one might be compelled to abandon them?

The Galilei-Newton transformation contains the hypothesis (hitherto not recognised as such) that there can be no finite velocity g which has the character of a universal constant. In other words, up to the present, the following was tacitly assumed to be true:

If an observer attached to a system S measure the velocity of propagation (v) of a certain effect, then an observer attached to a second system S', which is in motion relatively to S, would find a different value for the velocity of propagation of the same effect as measured from his system; and this applies for every finite value of the velocity v.

This assumption is, however, not confirmed in the case of the velocity of light *in vacuo* according to the experiments hitherto performed. To all appearances this velocity plays the part of a universal constant in physical nature; every observer finds the same value for the velocity of light, irrespective of his state of motion.

By taking due account of this experimental fact, and *combining* it with the postulate that systems which are in uniform rectilinear motion with regard to one another are to be equivalent, one arrives at the special relativity-theory postulated by Einstein, which is mathematically represented by the above group of equations of transformation of Lorentz-Einstein. This has in its turn led to the discovery of the remarkable fact that the conception of simultaneity upon which all time-measurements are founded, is devoid of meaning. It is possible to select a suitable time-coordinate in such a way that a time-measurement enters into physical laws in exactly the same manner as regards its significance as a space-measurement (i.e. they are fully equivalent symbolically), and has likewise a definite coordinate direction; a fact to which Minkowski in particular called attention.

The new form of the equations of transformation by no means exhausts the whole effect of the special principle of relativity upon classical mechanics. The change which it brought about in the conception of mass was almost still more marked. The experimental facts, which were mainly instrumental in bringing about this change, are those exhibited in the behaviour of electrons in motion.

Although it had been clearly recognised that electrons are freely moving energy-particles without associated material carriers, nevertheless they revealed in their own changes of motion all the characteristics of inertial matter in motion. But this inertia was not found to be a scalar quantity such as that of matter had always been considered to be hitherto. The special theory of relativity taught us that the conception of *apparent electromagnetic mass*, which had been deduced from the observations of electrons, was only a *particular* case of the inertia which is common to *all* forms of energy.

A hollow space, enclosed by reflecting walls of no mass and filled with *radiation*—(cavity-radiation)—when set in motion reveals the properties of inert mass, according to Maxwell's theory, on account of the enclosed energy of radiation. The fact that measurable deviations from the behaviour demanded by classical mechanics first exposed themselves in these inertial phenomena (of electrons) is above all due to the circumstance that velocities, that do not greatly differ from that of light, first made their appearance in these motions of the electrons. All the laws of mechanics have hitherto been tested only by motions in comparison with which the velocity of light was, practically speaking, infinitely great. The Lorentz-Einstein equations of transformation, however, show that motions approximating to the velocity of light take place in a manner quite different from that required by the laws of classical mechanics.

The discovery of the relativity of the inertia of energy created entirely new starting-points for erecting the structure of mechanics. Classical mechanics regards the inertial mass of a body as an absolute, invariable, characteristic quantity. The special theory of relativity, it is true, makes no mention of the inertial mass associated with *matter*. But as every kind of matter probably contains an enormous amount of latent energy, its inertia is composed of two components: (1) the inertia of the actual mass, (2) the inertia of the associated quantity of energy. The latter is not a scalar quantity; so that, in consequence, the sum of the two, as observed and measured by us, is also not a scalar. According to the present view of the general theory of relativity, all inertia of matter consists only of the inertia of the latent energy in it; in this case, everything that we know of the inertia of energy holds without exception for the inertia of *matter*.

Classical mechanics is, however, founded upon a conception of mass in which the inertial mass of a body is an attribute assumed to be peculiar to matter, and independent of all other physical conditions. The discovery of the inertia of energy, however, made it compulsory to arrange these phenomena into the scheme of mechanics.

The special theory of relativity, being confined to systems which are connected by the Lorentz-Einstein equations of transformation (*i.e. are in uniform rectilinear motion with regard to one another*), *is not equipped with the means of bringing this task to a satisfactory conclusion.*

The reason for this statement will become clear from the following example. One of the fundamental facts of mechanics is the equality of the inertial and gravitational mass of a body. It is on the supposition that this is true that we determine the mass of a body by measuring its weight. The *weight* of a body is, however, only defined *with reference to a gravitational field*: in our case, with reference to the earth. The idea of *inertial mass* of a body is, however, introduced as an attribute of matter *without any reference whatsoever to physical conditions external to the body*. How does the mysterious coincidence in the values of the inertial and gravitational mass of a body come about?

Nor does the special theory of relativity provide an answer to this question; for this purpose a theory of gravitational phenomena, a theory of gravitation, is required.

Furthermore, the special theory of relativity does not even preserve the equality in the values of inertia and gravitational mass; a fact which is to be reckoned amongst the most firmly established facts in the whole of physics. For, although the special theory of relativity makes allowance for an *inertia* of energy, it makes none for a *gravitation* of energy. Consequently, a body which absorbs energy in any way will register a gain of inertia but not of weight, thereby transgressing the principle of the equality of inertial and gravitational mass. The special theory of relativity can therefore be regarded only as a *stepping-stone* to a more general principle, which will enable us to take up a satisfactory position with regard to these various facts of experience *in toto*.

This is the point where Einstein's researches towards establishing a general theory of relativity set in. He has discovered that, by extending the application of the relativity-principle to accelerated motions, and by introducing gravitational phenomena into the consideration of the fundamental principles of mechanics, a new foundation for mechanics is made possible, in which all the difficulties occurring up to the present are solved. Although this theory represents a consistent development of the knowledge gathered by means of the special theory of relativity, it is so deeply rooted in the substructure of our principles of knowing, in their application to physical phenomena, that it is possible thoroughly to grasp the new theory only by clearly understanding its attitude toward these guiding lines provided by the theory of knowledge.

I shall, therefore, commence the account of his theory by discussing two general postulates, which should be fulfilled by every physical law, but neither of which is satisfied in classical mechanics: whereas their strict fulfilment is a characteristic feature of the new theory.

§ 2

TWO FUNDAMENTAL POSTULATES IN THE MATHEMATICAL FORMULATION OF PHYSICAL LAWS

NEWTON had established the simple and fruitful law that two bodies, even when they are not visibly connected with one another, as in the case of the heavenly bodies, exert a mutual influence, attracting one another with a force directly proportional to the product of their masses, and inversely proportional to the square of the distance between them. But Huygens and Leibniz refused to acknowledge the validity of this law, on the ground that it did not satisfy a fundamental condition to which every physical law is subject, viz. that of *continuity* (continuity in the transmission of force, action "by contact" in contradistinction to action "at a distance"). How were two bodies to exert an influence upon one another without a medium between them to transmit the action? The demand for a satisfactory answer to this question became in fact so imperative that finally, in order to satisfy it, the existence of a substance which pervaded the whole of cosmic space and permeated all matter—the "luminiferous ether"—was assumed, although this substance seemed to be condemned to remain intangible and invisible (i.e. imperceptible to the senses for all time) and had to be endowed with all sorts of contradictory properties. In the course of time, however, there arose in opposition to such assumptions the more and more definite demand that, *in the formulation of physical laws, only those things were to be regarded as being in causal connection which were capable of being actually observed*: a demand which doubtless originates from the same instinct in the search for knowledge as that of action "by contact," and which really gives the law of causality the true character of an *empirical* law, i.e. one of actual *experience.*

The consistent fulfilment of these two postulates combined together is, I believe, the mainspring of Einstein's method of investigation; this imbues his results with their far-reaching importance in the construction of a physical picture of the world. In this respect his endeavours will probably not encounter any opposition in the matter of principle on the part of scientists. *For both postulates*—(1) *that of continuity and* (2) *that of causal relationship between only such things as lie within the realm of observation*—are of an *inherent* nature, i.e. contained in the very nature of the problem. The only question that might be raised is whether it is *expedient* to abandon such useful working hypotheses as "forces at a distance."

The principle of continuity requires that all physical laws allow of formulation as differential laws, i.e. physical laws must be expressible in a form such that the physical state at any point is completely determined by that of the point in its immediate neighbourhood. Consequently the distances between points, which are at *finite* distances from one another, must not occur in these laws, but only those between points infinitely near to one another. The law of attraction of Newton given above, inasmuch as it involves "action at a distance," disobeys the first postulate.

The second postulate, that of a stricter form of expression for causality in its occurrence in physical laws, contains *the principle of the relativity of all motions* as a special case. It denotes the application of the former to the fundamental ideas of mechanics. As a matter of fact, we *observe* the motions of bodies only *relatively* to one another, whereas classical mechanics from the time of Newton onwards utilises the idea of *absolute* motion of a body in space. Einstein is the first to succeed in divesting mechanics entirely of these unnatural notions.

The rigorous application of the principles of continuity and relativity in their general form penetrates deeply into the problem of the mathematical formulation of physical laws. It will therefore be essential at the outset to enter into a consideration of the principles involved in the latter process.

§ 3

CONCERNING THE FULFILMENT OF THE TWO POSTULATES

A PHYSICAL law is clothed in mathematical language by setting up a formula. This comprises, and represents in the form of an equation, all measurements which numerically describe the event in question. We make use of such formulae not only in cases in which we have the means of checking the results of our calculations at any moment actually at our disposal: but also when the corresponding measurements cannot really be carried out *in practice* but have to be *imagined*, i.e. only take place in our minds: e.g. when we speak of the distance of the moon from the earth, and express it in metres, as if it were really possible to measure it by applying a metre-rule end to end.

By means of this expedient of analysis we have extended the range of exact scientific research far beyond the limits of measurement actually accessible in practice, both in the matter of immeasurably large, as well as in that of immeasurably small, quantities.

We have at the same time thereby created a symbolical method of representation, which expresses the events as being dependent on measurements of various kinds, e.g. time- and space-measurements, but unhampered by accidental limitations pronouncedly anthropomorphic in character. The discovery of suitable mathematical terms, which can be inserted in a formula as symbols for definite physical magnitudes of measurements, such as e.g. *length* of a rod, *volume* of a cube, etc., in order to shift the responsibility, as it were, for all further deductions upon analysis, is one of the fundamental problems of theoretical physics and *is intimately connected with the two postulates* enunciated in § 2.

To realise this fully, we must revert to the foundations of geometry, and analyse them from the point of view adopted by Helmholtz in various essays, and by Riemann in his inaugural dissertation of 1854: "On the hypotheses which lie at the bases of geometry." Riemann points almost prophetically to the path now taken by Einstein.

(*a*) THE LINE-ELEMENT IN THE THREE-DIMENSIONAL MANIFOLD OF POINTS IN SPACE, EXPRESSED IN A FORM COMPATIBLE WITH THE TWO POSTULATES.

Every point in space can be singly and unambiguously defined by the three numbers x_1, x_2, x_3, which may be regarded as the coordinates of a rectangular system of coordinates, and which distinguish it from all other points; a continuous variation of these three numbers enables us to specify every single point of space in turn. The assemblage of points in space represents, in Riemann's notation, "a multiply extended magnitude" (an n-fold manifoldness or manifold) between the single elements (points) of which a *continuous* transition is possible. We are familiar with diverse *continuous* manifolds, e.g. the system of colours, of tones and various others. A feature which is common to all of them is that, in order to specify a single element out of the entire manifold (to define a *particular* point, a *particular* colour, or a *particular* tone), a characteristic number of magnitude-determinations, i.e. coordinates, is required: this characteristic number is called the *dimensions* of the respective manifold. Its value is three for space, two for a plane, one for a line. The system of colours is a continuous manifold of the dimension three, corresponding to the three "primary" colours red, green and violet, by mixing which in due proportions *every* colour can be produced.

But the assumption of continuity for the transition from one element to another in the same manifold, and the determination of the dimensions of the latter, does not give us any information about the possibility of comparing limited parts of the same manifold with one another, e.g. about the possibility of comparing two tones with one another or two single colours; i.e. nothing has yet been stated about the metric relations (measure-conditions) of the manifold, about the nature of the scale, according to which measurements can be undertaken within the manifold. In order to be able to do this, we must allow experience to give us the facts from which to establish the metric (measure-) laws which hold for each particular manifold (space-points, colours, tones) under various physical conditions; these metric laws will be different according to the set of empirical facts chosen for this purpose*.

In the case of the manifold of space-points experience has taught us that *finite rigid* point-systems can be freely moved in space without altering their form or dimensions; the conception of "congruence" which has been derived from this fact, has become a vital factor for a measure-determination †. It sets us the problem of building up an expression from the numbers x_1, x_2, x_3, and y_1, y_2, y_3, denoting two definite points in space, which can be regarded as a measure of their fixed or invariable distance from one another and as such can be introduced into physical laws.

The equations of physical laws, which—in order to fulfil the conditions of continuity—must be differential laws, only contain the distances ds, of infinitely near points, so-called *line-elements*. We must therefore enquire whether our two postulates of § 2 have any influence upon the analytical expression for the *line-element ds* and, if so, which expression for the latter is compatible with both. Riemann demands of a line-element in the first place that it can be *compared* in respect to its *length* with every other line-element independent of its position and direction. This is a distinguishing characteristic of the metric conditions ("measure relations") prevalent in space; this peculiarity does not exist, for instance, in the manifold of tones or in that of colours (*vide* Note 2). Riemann formulates this condition in the words, "that lines must have a length independent of their position and that every line is to be measurable by means of any other." He then discovers that: if x_1, x_2, x_3 and $x_1 + dx_1$, $x_2 + dx_2$, $x_3 + dx_3$ respectively denote two infinitely near points in space and if the continuously variable numbers x_1, x_2, x_3 are *any* coordinates whatsoever (not e.g. necessarily rectilinear), then the square root of an always positive, integral, homogeneous function of the second degree in the differentials dx_1, dx_2, dx_3 has all the properties ‡ which the line-element must exhibit. We thus find that:

$$ds = \sqrt{g_{11}dx_1^2 + g_{12}dx_1dx_2 + \dots + g_{33}dx_3^2},$$

* *Vide* Note 2. † *Vide* Note 3. ‡ *Vide* Note 4.

in which the coefficients $g_{\mu\nu}$ are continuous functions of the three variables x_1, x_2, x_3, gives us an expression for the line-element at the point x_1, x_2, x_3.

No assumptions whatsoever are contained in this expression concerning the question as to the coordinates of the three variables that are to be used, whether they are to be Cartesian, polar, or curvilinear (Gaussian) (*vide* page 32), i.e. about the nature of the co-ordination of space. If it be in particular demanded that each point can be defined by means of Cartesian coordinates x, y, z the line-element assumes the form

$$ds = \sqrt{dx^2 + dy^2 + dz^2},$$

for this special case.

This latter expression has hitherto always been introduced into all physical laws, as it allows the use of Euclidean geometry for all space-measurements. But this particular assumption contains the hypothesis, as Helmholtz has shown in a detailed discussion, that *finite* rigid point-systems, i.e. *finite* fixed distances, are capable of unrestrained motion in space, and can be made (by superposition) to coincide with other (congruent) point-systems. With respect to the postulate of *continuity*, this hypothesis seems inconsistent, in so far as it introduces implicit statements about *finite* distances into purely differential laws, in which only *line-elements* occur; but it does not *contradict* the postulate.

The postulate of the relativity of all motion takes a different stand towards the possibility of giving the line-element the Euclidean form in particular*.

According to the principle of the relativity of all motions, all systems, which come about owing to relative motions of bodies towards one another, may be regarded as fully equivalent. The laws of physics must therefore preserve their form in passing from one such system to another; i.e. the transformation-formulae of the variables which perform this transition to another system, must not alter the analytical expression for the physical law under consideration.

* Strictly speaking I should at this juncture state in anticipation that the above investigations can manifestly also be so generalised as to be valid for the four-dimensional *space-time* manifold, in which all events actually take place, and that the transformation-formulae apply to the four variables of this manifold. In these general remarks the neglect of the fourth dimension is of no importance. This statement will be justified later in § 3 (*b*).

As we have to reckon with all possible relative motions of bodies with respect to one another, the *general* principle of relativity requires that the physical laws, and thereby also the line-element which occurs in them, preserve their form for *every arbitrary* transformation of the variables. This condition is fully satisfied by the line-element:

$$ds = \sqrt{g_{11}dx_1^2 + g_{12}dx_1dx_2 + \ldots + g_{33}dx_3^2},$$

in which no restrictive reservations of any description are made as to what the coordinates x_1, x_2, x_3 are to signify. The Euclidean line-element

$$ds = \sqrt{dx^2 + dy^2 + dz^2},$$

on the other hand, preserves its form *only* for transformations of the *special* theory of relativity (*vide* § 1 and Note 1), which confine themselves to systems moving *uniformly* and rectilinearly. Moreover, experience teaches us daily that bodies continually move with *accelerated motion* towards one another *as a result of their mutual gravitational influences*. Consequently the postulate of a principle of relativity for *all* motions is not to be brought into agreement with the limitation imposed by adopting the Euclidean element of arc for the differential laws of physics.

The choice of the expression $ds^2 = \sum_1^3 g_{\mu\nu}dx_\mu dx_\nu$ to represent the line-element in physical laws is, in spite of its very general character, still to be regarded as a hypothesis, as Riemann has already pointed out. For there are other functions of the differentials dx_1, dx_2, dx_3—such as e.g. the fourth root of a homogeneous differential expression of the fourth degree in these variables—which could provide a measure for the length of the line-element (*vide* Note 5). But at present there is no ground for abandoning the simplest general expression for the line-element, viz. that of the second degree, and adopting more complicated functions. Within the range (of fulfilment) of the two postulates, which we have imposed upon every description of physical events, the former expression for ds satisfies all requirements. Nevertheless, it must never be forgotten that the choice of an analytical expression for the line-element always contains a hypothetical factor; and it is the duty of the physicist to remain fully conscious of this fact at all times, without being in any way prejudiced. It is for this reason that Riemann closes his essay with the following remarks, which impress one particularly with their great importance for the present time.

"The question of the validity of the hypotheses of geometry in the infinitely small is bound up with the question of the ground of the metric relations of space. In this question, which we may still regard as belonging to the doctrine of space, is found the application of the remark made above; that in a discrete* manifold, the principle or character of its metric relations is already given in the notion of the manifold, whereas in a continuous manifold this ground has to be found elsewhere, i.e. has to come from outside. Either, therefore, the reality which underlies space must form a discrete manifold, or we must seek the ground of its metric relations (measure-conditions) outside it, in binding forces which act upon it.

A decisive answer to these questions can be obtained only by starting from the conception of phenomena which has hitherto been justified by experience, to which Newton laid the foundation, and then making in this conception the successive changes required by facts which admit of no explanation on the old theory; researches of this kind, which commence with general notions, cannot be other than useful in preventing the work from being hampered by too narrow views, and in keeping progress in the knowledge of the interconnections of things from being checked by traditional prejudices.

This carries us over into the sphere of another science, that of physics, into which the character and purpose of the present discussion will not allow us to enter."

That is to say: according to Riemann's view these questions are to be solved by starting from Newton's view of physical phenomena, and compelled by facts which do not allow of any explanation by it, gradually remoulding it. This is what Einstein has done. The "binding forces," to which Riemann points, will be found again in Einstein's theory. As we shall see in the fifth chapter, Einstein's theory of gravitation is based upon the view that the gravitational forces are the "binding forces," i.e. they represent the "inner ground" of the metric conditions (measure-relations) of the space-time-manifold.

* *Vide* Note 6.

(*b*) THE LINE-ELEMENT IN THE FOUR-DIMENSIONAL MANIFOLD OF SPACE-TIME POINTS, EXPRESSED IN A FORM COMPATIBLE WITH THE TWO POSTULATES.

The measure-conditions, which we were to take as a basis for the formulation of physical laws, could have been treated immediately in connection with the four-dimensional manifold of space-time points, seeing that, according to the special theory of relativity, time-measurements enter into physical laws exactly like space-measurements. Nevertheless, I wish to treat time-measurements separately; for one reason, that it is just this result of the relativity-theory which has experienced the greatest opposition at the hands of supporters of classical mechanics: and for another that classical mechanics is also obliged to establish certain conditions about time-measurement.

In Galilei's law of inertia, a body which is not subject to external influences continues to move with uniform motion in a straight line. Two determining elements are lacking, viz. the reference of the motion to a definite system of coordinates, and a definite time-measure. Without a time-measure one cannot speak of a *uniform* velocity.

Following a suggestion by C. Neumann*, the law of inertia has itself been adduced to give a definition of a time-measure in the form: "Two material points, both left to themselves, move in such a way that equal lengths of path of the one correspond to equal lengths of path of the other." On this principle, into which time-measure does not enter explicitly, we can define "equal intervals of time as such, within which a point, when left to itself, traverses equal lengths of path."

This is the attitude which was also taken up by L. Lange, H. Seeliger and others, in later researches. Maxwell selected this definition too (in *Matter and Motion*). On the other hand, H. Streintz† (following Poisson and d'Alembert) has demanded the disconnection and independence of the time-measure from the law of inertia, on the ground that the roots of the time-concept have a deeper and more general foundation than the law of inertia.

* *Vide* Note 7. † *Vide* Note 8.

According to his opinion, *every* physical event, which can be made to take place again under exactly the same conditions, can serve for the determination of a time-measure, inasmuch as every *identical* event must claim precisely the same duration of time; otherwise, an ordered description of physical events would be out of the question. In point of fact, the clock is constructed on this principle. It is this principle which enables an observer to undertake a time-measurement *at least for his place of observation.* The reduction of time-measurements to a dependence upon the law of inertia, on the other hand, leads to an unobjectionable *definition of equal lengths of time*; but the *measurement* of the equal paths traversed by uniformly moving bodies, and the *establishment* of a unit of time involved therein, are only then possible for a place of observation, when the observer and the moving body are in constant connection, e.g. by light-signals. It cannot however be straightway assumed that two observers, who are in rectilinear motion relatively to one another, and therefore, according to the law of inertia, equivalent as reference systems, would in this manner gain identical results in their time-measurements.—Poisson's idea thus leads to a satisfactory time-measurement *for a given place of observation* itself; i.e. in a certain sense it allows the construction of a clock for that place. But it does not broach the question of the time-relations of *different* places with *one another* at all; whereas Neumann's suggestion leads directly to those questions which have been a centre of discussion since Einstein's enunciation of the relativity-principle.

In the endeavour to reduce classical mechanics to as small a number of principles as possible, in perfect agreement with one another, writers resorted to *ideal-constructions* and *imaginary experiments*. It never occurred to anyone that the use of a light-signal as a means of connection between the moving-body and the observer, which is necessary in practice in order to determine *simultaneity*, might affect the final result, i.e. of time-measurements in different systems. But according to Einstein the truth of just this conjecture cannot be ignored; because the conception of *simultaneity*, upon which all time-measurements are based, has no meaning in itself, i.e. absolutely (*vide* Note 9). Consequently the accepted conventions of classical mechanics about the measurement of time are insufficient.

That such a fundamental revision of the assumptions made regarding time-measurements became necessary only after so great a lapse of time, is to be explained by the fact that even the velocities which occur in astronomy are so small, in comparison with the velocity of light, that no serious discrepancies could arise between theory and observation. So it occurred that the weaknesses of the theory—in particular, those due to the motional relations of various systems to one another—did not come to light. It was, above all, not recognised that the equations of transformation of the relativity-theory of Newton-Galilei, which express the relations between the coordinates of two systems in uniform rectilinear motion with respect to one another (i.e. of two systems mechanically equivalent: in which therefore time-measurements are assumed to be fully independent of one another), contain hypotheses. The relativity-theory of Einstein first disclosed them. This will presently become more evident.

In principle, the following question might have been proposed long before the discovery of electrodynamical phenomena: how are the measurements x, y, z, t and x', y', z', t' in two different coordinate systems, which are in uniform rectilinear motion relatively, to be related to one another in general; i.e. in what way are the x, y, z, t expressible in terms of x', y', z', t' and q, the relative velocity of the two systems?—a question to which Neumann's proposal for a time-measurement directly points. From perfectly general considerations, arising from certain fundamental ideas about motions, one would have arrived at equations of transformation of a much more general character than that of the relativity-principle of Newton-Galilei, in which t' is always put equal to t (*vide* Note 10). In these supposed general equations of transformation, one magnitude would have claimed special attention. Whenever any effect propagates itself with the velocity v in a certain system, then it will in general propagate itself, when referred to another system which is in motion relatively to the first, with a velocity in general differing from v, i.e. $v' \neq v$. But, according to Frank and Rothe, there is always one unique velocity for *every* system, which preserves its value, independent of the motion of the system. It had hitherto been tacitly assumed that only infinite velocity possessed this special property. In this particular case the general equations of transformation degenerate into those of Newton and Galilei (*vide* Note 1). If the question as to whether there is a *finite* velocity, which reveals this special property, had been left *open*, one would at least have remained aware of the hypothetical element in this assumption, and the result of Michelson and Morley's experiment—viz. that the velocity of light actually reveals this property—as well as Einstein's deductions therefrom concerning the measurement of time, would not have been felt to be such an arbitrary encroachment upon mechanics.

The character of universal significance possessed by the velocity of light must be accepted as an established fact.

This shows that the assumptions hitherto made about the time-measurements, upon which mechanics was supposed to be based, are not compatible with the equations of transformation of the relativity-principle of Galilei and Newton and at the same time with the fact of the constancy of the velocity of light: we are thus compelled to call into action the views first developed by Einstein, which take into account the relativity of time-measurements.

The details of the effects, which result from the relativity of the time-concept, have so often been discussed in recent years that it is only possible to repeat what has already often been said. An essential point is the recognition of the fact, that time-measurements enter into physical laws as in every sense equivalent with space-measurements, and have similarly their corresponding co-ordinate-direction. Space and time therefore represent a *homogeneous* manifold of "four" dimensions with *homogeneous measure-relations* (*vide* Note 11). Consequently, to be consistent, we must apply the arguments of the preceding § 3 (*a*) about the measure-relations to the four-dimensional space-time-manifold; and, in view of the two fundamental postulates (1) of continuity and (2) of relativity, and including the time-measurement as the fourth dimension, we must select for our line-element the expression:

$$ds^2 = g_{11}dx_1^2 + g_{12}dx_1dx_2 + \dots + g_{34}dx_3dx_4 + g_{44}dx_4^2,$$

in which the $g_{\mu\nu}$ (μ, $\nu = 1, 2, 3, 4$) are functions of the variables x_1, x_2, x_3, x_4.

Hitherto we have been led to adopt this much more general

attitude towards the questions of the metric laws involved in physical formulae merely by the desire not to introduce, from the very outset, more assumptions into the formulations of physical laws than are compatible with both postulates, and to bring about a deeper appreciation of the points of view, to which the *special* theory of relativity has led us.

We can briefly summarise by saying: the adoption of *Euclidean* metric-conditions (measure-relations) is *compatible* with the postulate of *continuity*; though the special assumptions thereby involved appear as restrictive or limiting hypotheses, which need not be made. But the second postulate, the reduction of *all* motions to relative motions, *compels* us to abandon the Euclidean measure-determination (cf. bottom of page 14). A description of the difficulties still remaining in mechanics will make this step clear.

§ 4

THE DIFFICULTIES IN THE PRINCIPLES OF CLASSICAL MECHANICS

THE foundations of classical mechanics cannot be exhaustively described in a narrow space. I can only bring the unfavourable side of the theory into prominent view for the present purpose, without being able to do justice to its great achievements in the past. All doubts about classical mechanics set in at the very commencement with the formulation of the law which Newton places at its head, the formulation of the law of inertia.

As has already been emphasised on page 16, the assertion that a point-mass which is left to itself moves with uniform velocity in a straight line, omits all reference to a definite coordinate system. An insurmountable difficulty here arises: Nature gives us actually no coordinate system, with reference to which a uniform rectilinear motion would be possible. For as soon as we connect a coordinate system with any body such as the earth, sun or any other body—and this alone gives it a physical meaning—the first condition of the law of inertia (viz. freedom from external influences) is no longer fulfilled, on account of the mutual gravitational effects of the bodies. One must accordingly either assign to the motion of the body a meaning in itself, i.e. grant the existence of motions relative to "absolute" space, or have recourse to mental experiments by following the example of C. Neumann and introducing a hypothetical body Alpha, relative to which a system of axes is defined, and with reference to which the law of inertia is to hold (Inertial system, *vide* Note 12). The alternative with which one is faced is highly unsatisfactory. The introduction of absolute space gives rise to the oft-discussed conceptual difficulties which have gnawed at the foundations of Newton's mechanics. The introduction of the system of reference Alpha certainly takes the relativity of motions so far into account, that all systems in uniform motion relative to an Alpha-system are established as *equivalent* from the very outset, but we can affirm with certainty that there is no such thing as a visible Alpha-system, and that we shall never succeed in arriving at a final determination of such a system. (It will, at most, be possible, by means of progressively taking account of the influences of constellations upon the solar system and upon one another, to approximate to a system of coordinates, which could play the part of such an inertial system *with a sufficient degree of accuracy.*) As a result of this objection, the founder of the view himself, C. Neumann, admits that it will always be somewhat unsatisfactory and enigmatical, and that mechanics, based on this principle, would indeed be a very peculiar theory.

It therefore seems quite natural that E. Mach (*vide* Note 13) should be led to propose that the law of inertia be so formulated that its relations to the stellar bodies are directly apparent. "Instead of saying that the direction and speed of a mass u remains constant in space, we can make use of the expression that the mean acceleration of the mass μ relative to the masses m, m', m'' ... at distances r, r', r'' ..., respectively, is zero or

$$\frac{d^2}{dt^2} \cdot \frac{\Sigma mr}{\Sigma m} = 0.$$

The latter expression is equivalent to the former statement, as soon as a sufficient number and sufficiently great and extensive masses are taken into consideration...." This formulation cannot satisfy us. For, in addition to a certain requisite accuracy, the character of a "contact" law is lacking, so that its promotion to the rank of a fundamental law (in place of the law of inertia) is quite out of the question.

The inner ground of these difficulties is without doubt to be found in an insufficient connection between fundamental principles and observation. As a matter of actual fact, we only *observe* the motions of bodies *relatively* to one another, and these are *never* absolutely rectilinear nor uniform. *Pure inertial motion is thus a conception deduced by abstraction from a mental experiment—a mere fiction.*

As fruitful and unavoidable as a mental experiment may often be, there is the ever present danger that an abstraction which has been carried unduly far loses sight of the physical contents of its underlying notions. And so it is in this case. If there is no meaning for our understanding in talking of the "motion of a body" in space, as long as there is only this *one* body present, is there any meaning in granting the body attributes such as *inertial mass*, which only arise from our observation of *several* bodies, moving *relatively* to one another? If not, we can attach only a relative meaning to the conception "inertial mass of a body."

The results of the special theory of relativity entirely unhinged our view of the inertia of matter, for they robbed the theorem concerning the equality of inertial and gravitational mass of its strict validity. A body was now to have an *inertial* mass varying with its contained internal energy*, without its *gravitational* mass being altered. But the mass of a body had always been ascertained from its weight, without any inconsistencies manifesting themselves (*vide* Note 15).

A difficulty of such a fundamental character could come to light only owing to the theorem of the equality of inertial and gravitational mass not being sufficiently interwoven with the underlying principles of mechanics, and because the same importance had not been accorded to gravitational phenomena as to inertial phenomena, which, judged from the standpoint of experience, must be claimed. Gravitation, as a force acting at a distance, is, on the contrary, introduced only as a special force for a limited range of phenomena: and the surprising fact of the equality of inertial and gravitational mass, valid at all times and in all places, receives no further attention. *One must therefore substitute for the law of inertia a fundamental law which comprises inertial and gravitational phenomena. This can be brought about by a consistent application of the principle of the relativity of all motions, as Einstein has recognised. This is therefore the circumstance chosen by Einstein as a nucleus about which to weave his developments.*

The theorem of the equality of inertial and gravitational mass, which reflects the intimate connection between inertial and gravitational phenomena, may be illuminated from another point of vantage, and thereby disclose *its close relationship* (*vide* page 7) *to the general principle of relativity.*

However much the notion of "absolute space" repelled Newton, he nevertheless believed he had a strong argument, in support of the existence of absolute space, in the phenomenon of centrifugal forces. When a body rotates, centrifugal forces make their appear-

* *Vide* Note 14.

ance. Their presence in a body alone, *without any other visible body being present*, enables one to demonstrate the fact that it is in rotation. Even if the earth were perpetually enveloped in an opaque sheet of cloud, one would be able to establish its daily rotation about its axis by means of Foucault's pendulum-experiment. This peculiarity of rotations led Newton to conclude that absolute motions exist. From the *purely kinematical* point of view, however, the rotation of the earth is not to be distinguished in any way from a translation; in this case, too, we *observe* only the *relative* motions of bodies, and might just as well imagine that all bodies in the universe revolve around the earth. E. Mach has in fact affirmed that both events are equivalent, not only kinematically, but also *dynamically*: it must, however, then be assumed that the *centrifugal forces*, which are observed *at the surface of the earth*, would also arise, equal in quantity and similar in their manifestations, from the *gravitational effect* of all bodies in their entirety; if these revolved around the supposedly *fixed* earth (*vide* Note 16).

The justification for this view, which in the first place arises out of the kinematical standpoint, is in the main to be sought in the fact, derived from experience, that inertial and gravitational mass are equal. According to the conceptions, which have hitherto prevailed, the centrifugal forces are called into play by the inertia of the rotating body (or rather by the inertia of the separate points of mass, which continually strive to follow the bent of their inertia, and therefore express the tendency to fly off at a tangent to the path in which they are constrained to move). The field of centrifugal forces is therefore an *inertial field* (*vide* Note 17). The possibility of *regarding* it equally well as a *gravitational field*—and we do that, as soon as we also *assert* the relativity of rotations dynamically: for we must then assume that the whole of the masses describing paths about the (supposed) fixed body *induce* the so-called centrifugal forces by means of their gravitational action—is founded on the equality of inertial and gravitational mass, a fact which Eötvös has established with extraordinary precision by making use of the centrifugal forces of the rotating earth (*vide* Note 18). *From these considerations one recognises how a general principle of the relativity of all motions simultaneously implies a theory of gravitational fields.*

From these remarks one inevitably gains the impression that a construction of mechanics upon an entirely new basis is an absolute necessity. There is no hope of a satisfactory formulation of the law of inertia without taking into account the *relativity of all motions*, and hence just as little hope of banishing the unwelcome conception of absolute motion out of mechanics; moreover the discovery of the *inertia of energy* has taught us facts which refuse to fit into the existing system, and necessitate a revision of the foundations of mechanics. The condition which must be imposed at the very outset (cf. page 8) is: Elimination of all actions which are supposed to take place "at a distance" and of all quantities which are not capable of direct observation, out of the fundamental laws; i.e. **the setting-up of a differential equation which comprises the motion of a body under the influence of *both* inertia *and* gravity and symbolically expresses the relativity of all motions.** This condition is completely satisfied by Einstein's theory of gravitation and the general theory of relativity. The sacrifice, which we have to make in accepting them, is to renounce the hypothesis, which is certainly deeply rooted, that all physical events take place in space which is "equipped" with the axioms of Euclidean geometry. For the postulate of general relativity, which also applies to accelerated motions, demands that the fundamental laws be independent of the particular choice of the coordinates of reference. But the Euclidean line-element does not preserve its form after any arbitrary change of the coordinates of reference. We have, therefore, to substitute in its place the general line-element:

$$ds^2 = \sum_{1}^{4} g_{\mu\nu} dx_\mu dx_\nu.$$

Whereas, then, the postulate of continuity (cf. page 9) only seemed to render it *advisable* not to introduce the narrowing assumptions of the Euclidean determination of measure, the principle of general relativity no longer leaves us any choice.

The reason for so emphasising the latter principle—as indeed also the postulate that only observable quantities are to occur in physical laws—is not to be sought in any requirement of a merely formal nature, but rather in an endeavour to invest the principle of causality with the authority of a law which holds good in the world of actual physical experience. One must above all avoid introducing into physical laws, side by side with observable quantities, hypotheses which are purely fictitious in character, as e.g. the *space* of Newton's mechanics. Otherwise the principle of causality would not give us any real information about the causes and effects, i.e. the causal relations of the contents of *direct* experience; which is presumably the aim of every physical description of natural phenomena. The fact that the postulate of the relativity of all motions is so deeply rooted in the *foundations of all our knowledge* gives us the standpoint from which its true merits are to be judged (*vide* Note 19).

§ 5

Einstein's Theory of Gravitation

(*a*) THE FUNDAMENTAL LAW OF MOTION AND THE PRINCIPLE OF EQUIVALENCE OF THE NEW THEORY

ACCORDING to the preceding discussion, the law of inertia of mechanics must be replaced by an entirely different one, viz. by a differential law, which in the first place describes the motion of a point-mass under the influence of both inertia *and* gravity, and which, secondly, always preserves the same form, irrespective of the system of coordinates to which it be referred, so that no system of coordinates enjoys a preference to any other. (The first condition arises from the necessity of ascribing the same importance to gravitational phenomena as to inertial phenomena in the new process of founding mechanics—the law must therefore also contain terms which denote the gravitational state of the field from point to point; the second condition is derived from the postulate of the relativity of all motion.)

A law of this kind exists in the special theory of relativity in the equation of motion of a single point, not subject to any external influence, in the form:

$$\delta\{\int ds\} = \delta\{\int \sqrt{-dx^2 - dy^2 - dz^2 + c^2dt^2}\} = 0.$$

According to this equation, the path of a point is the "shortest" or "straightest" line (*vide* Note 20)—i.e. the "straight line," if the line-element ds is Euclidean. If the principle of the shortest path, which is to be followed in actual motions, be elevated in this form to a general *differential law* for the motion in a *gravitational* field too, with due regard to the principle of the relativity of all motions, the new fundamental law must run as follows:

$$\delta\{\int ds\} = \delta\{\int \sqrt{g_{11}dx_1^2 + g_{12}dx_1dx_2 + \dots + g_{44}dx_4^2}\} = 0 \dots (1).$$

For only this form of the line-element remains unaltered (invariant) for arbitrary transformations of the x_1, x_2, x_3, x_4. The ten coefficients $g_{\mu\nu}$ which will in general be functions of the variables $x_1, \dots x_4$, must be able to be brought into such relationship to the gravitational field, in which the motion takes place, that they are determined by the field, and that the motion *described* by equation (1) coincides with the *observed* motion. This is actually possible. (The $g_{\mu\nu}$'s are the gravitational potentials of the new theory, i.e. they take over the part played by the one gravitational potential in Newton's theory, without, however, having the special properties, which according to our knowledge a potential has, in addition.)

Corresponding to the measure-relations of a space-time manifold based upon the line-element:

$$ds^2 = \sum_1^4 g_{\mu\nu} dx_\mu dx_\nu,$$

which is now placed at the foundation of mechanics by virtue of the relativity of all motions, the remaining physical laws must also be so formulated that they remain independent of the accidental choice of the variables. Before we enter into this more closely, the distinguishing features of the theory of gravitation characterised by equation (1) will be considered in greater detail.

The postulate of the new theory, that the laws of mechanics are only to contain statements about the relative motions of bodies, and that, in particular, the motion of a body under the action of the attraction of the remaining bodies is to be symbolically described by the formula:

$$\delta \left\{ \sqrt{\sum_1^4 g_{\mu\nu} dx_\mu dx_\nu} \right\} = 0,$$

is fulfilled in Einstein's theory by a physical hypothesis concerning the nature of gravitational phenomena, which he calls the *hypothesis* or *principle* (respectively) *of equivalence* (*vide* Note 21). This asserts the following:

*Any change, which an observer perceives in the passing of any event to be due to a gravitational field, would be perceived by him in exactly the same way, if the gravitational field were not present, provided that he —the observer—makes his system of reference move with the acceleration which was characteristic of the gravitation at his point of observation**.

For if the variables x, y, z, t in the equation of motion

$$\delta \{\int ds\} = \delta \left\{\int \sqrt{- dx^2 - dy^2 - dz^2 + c^2 dt^2}\right\} = 0$$

* This is discussed in an elementary way in a short essay "The Theory of Relativity" by the translator, which has been issued as a pamphlet by B. H. Blackwell, Broad Street, Oxford.

of a point-mass moving uniformly and rectilinearly (i.e. uninfluenced by gravity) be subjected to any transformation corresponding to the change of the x, y, z, t into the coordinates x_1, x_2, x_3, x_4 of a system of reference which has any accelerated motion whatsoever with regard to the initial system x, y, z, t; then, in general, coefficients $g_{\mu\nu}$ will occur in the transformed expression for ds, and will be functions of the new variables $x_1, \ldots x_4$, so that the transformed equation will be:

$$\delta \left\{\int \sqrt{g_{11}dx_1{}^2 + g_{12}dx_1 dx_2 + \ldots + g_{44}dx_4{}^2}\right\} = 0.$$

Taking into account the extended region of validity of this equation, one will be able to regard the $g_{\mu\nu}$ which *arise* from the accelerational transformation (*vide* Note 22) just as well, as due to the action of a gravitational field, which asserts its existence in effecting just these accelerations. *Gravitational problems thus resolve into the general science of motion of a relativity-theory of all motions.*

By thus accentuating the equivalence of gravitational and accelerational events, we raise the fundamental fact, that *all* bodies in the gravitational field of the earth fall with *equal* acceleration, to a fundamental assumption for a new theory of gravitational phenomena. This fact, in spite of its being reckoned amongst the most certain of those gathered from experience, has hitherto not been allotted any position whatsoever in the foundations of mechanics. On the contrary, the Galilean law of inertia makes an event which had never been actually observed (the uniform rectilinear motion of a body, which is not subject to external forces) function as the main-pillar amongst the fundamental laws of mechanics. This brought about the strange view that inertial and gravitational phenomena, which are probably not less intimately connected with one another than electric and magnetic phenomena, have nothing to do with one another. The phenomenon of inertia is placed at the head of classical mechanics as the fundamental property of matter, whereas gravitation is only, as it were, introduced by Newton's law as one of the many possible forces of nature. The remarkable fact of the equality of the inertial and gravitational mass of bodies only appears as an accidental coincidence.

Einstein's principle of equivalence assigns to this fact the rank to which it is entitled in the theory of motional phenomena. The new equation of motion (1) is intended to describe the relative motions of bodies with respect to one another under the influence of *both* inertia *and* gravity. The gravitational and inertial phenomena are amalgamated in the *one* principle that the motion take place in the geodetic line ($\delta \int ds = 0$). Since the element of arc

$$ds = \sqrt{\sum_1^4 g_{\mu\nu} dx_\mu dx_\nu}$$

preserves its form after any arbitrary transformation of the variables, all systems of reference are *equally* justified as such, i.e. there is none which is more important than any other.

The most important part of the problem, with which Einstein saw himself confronted, was the setting-up of differential equations for the gravitational potentials $g_{\mu\nu}$ of the new theory. With the help of these differential equations, the $g_{\mu\nu}$'s were to be unambiguously calculated (i.e. as single-valued functions) from the distribution of the quantities exciting the gravitational field; and the motion (e.g. of the planets) which was described, according to equation (1) by inserting these values for the $g_{\mu\nu}$'s, had to agree with the observed motion, if the theory was to hold true. In setting up these differential equations for the gravitational potentials $g_{\mu\nu}$ Einstein made use of hints gathered from Newton's theory, in which the factor which excites the field in Poisson's equation $\Delta\phi = -4\pi\rho$ for the Newtonian gravitational potential (viz. the factor represented by ρ, the density of mass in this equation) is put proportional to a *differential expression of the second order.* This circumstance prescribes, as it were, the method of building up these equations, taking for granted that they are to assume a form similar to that of Poisson's equation.

In conformity with the deepened meaning we have assigned to the mutual relation between inertia and gravity, as well as to the connectión between the inertia and latent energy of a body, we find that ten components of the quantity which determines the "*energetic*" state at any point of the field, and which was already introduced by the special theory of relativity as "*stress-energy-tensor*," duly make their appearance in place of the density of mass ρ, in Poisson's equation.

Concerning the differential expressions of the second order in the $g_{\mu\nu}$'s which are to correspond to the $\Delta\phi$ of Poisson's equation,

Riemann has shown the following: the measure-relations of a manifold based on the line-element

$$ds^2 = \sum_1^4 g_{\mu\nu} dx_\mu dx_\nu,$$

are in the first place determined by a differential expression of the fourth degree (the Riemann-Christoffel Tensor), which is independent of the arbitrary choice of the variables $x_1, \ldots x_4$ and from which all other differential expressions which are likewise independent of the arbitrary choice of the variables $x_1, \ldots x_4$ and only contain the $g_{\mu\nu}$'s and their derivatives, can be developed (by means of algebraical and differential operations). This differential expression leads unambiguously, i.e. in only one possible way, to ten differential expressions in the $g_{\mu\nu}$'s. And now, in order to arrive at the required differential equations, Einstein puts these ten differential expressions proportional to the ten components of the stress-energy-tensor, regarding the latter ten as the quantities exciting the field. He inserts the gravitational constant as the constant of gravitation. **These differential equations for the $g_{\mu\nu}$'s, together with the principle of motion given above, represent the fundamental laws of the new theory.** To the first order they, in point of fact, lead to those forms of motion, with which Newton's theory has familiarised us (*vide* Note 23). More than this, without requiring the addition of any further hypothesis, they mathematically account for the only phenomenon in the theory of planetary motion which could not be explained on the Newtonian theory, viz. the occurrence of the remainder-term in the expression for the motion of Mercury's perihelion.

Since the formulae of the new theory are based upon a space-time-manifold, the line-element of which has the general form

$$ds = \sqrt{\sum_1^4 g_{\mu\nu} dx_\mu dx_\nu},$$

all other physical laws, in order to bring the general theory of relativity to its logical conclusion, must receive a form which, in agreement with the new measure-conditions, must be independent of the arbitrary choice of the four variables x_1, x_2, x_3, x_4.

Mathematics has already performed the preliminary work for the solution of this problem in the calculus of absolute differentials; Einstein has elaborated them for his particular purposes (in his essay "Concerning the formal foundations of the general theory of relativity*"); Gauss invented the calculus of absolute differentials in order to study those properties of a surface (in the theory of surfaces) which are not affected by the position of the surface in space nor by inelastic continuous deformations of the surface (deformations without tearing), so that the value of the line-element does not alter at any point of the surface. As such properties depend upon the *inner* measure-relations of the surface only, one avoids referring, in the theory of surfaces, to the usual system of coordinates, i.e. one avoids reference to points which do not themselves lie on the surface. Instead of this, every point in the surface is fixed, by covering the surface with a net-work, consisting of two intersecting arbitrary systems of curves, in which each curve is characterised by a parameter; every point of the surface is then unambiguously, i.e. singly, defined by the two parameters of the two curves (one from each system) which pass through it, i.e. of which it is the point of intersection. According to this view of surfaces, a cylindrical envelope and a plane, for instance, are not to be regarded as different configurations: for each can be unfolded upon the other without stretching, and accordingly the same planimetry holds for both—a criterion that the inner measure-relations of these two manifolds are the same (*vide* Note 24). The general theory of relativity is based upon the same view; but now *not* as applied to the two-dimensional manifold of surfaces, but with respect to the *four-dimensional* space-time manifold. As the four space-time variables are devoid of all physical meaning, and are only to be regarded as four parameters, it will be natural to choose a representation of the physical laws, which provides us with differential laws which are independent of the chance choice of the x_1, x_2, x_3, x_4; this is what is done by the calculus of absolute differentials. The results of the preceding paragraphs, the far-reaching consequences of which can be fully recognised only by a detailed study of the mathematical developments involved, may be summarised as follows:

The calculus of absolute differentials of a space-time-manifold based upon the general line-element

$$ds^2 = \sum_1^4 g_{\mu\nu} dx_\mu dx_\nu,$$

* "Über die formalen Grundlagen der allgemeinen Relativitätstheorie," *Sitz. Ber. d. Kgl. Preuss. Akad. d. Wiss.* XLI. 1916, S. 1080.

enables us to obtain for every law of the special theory of relativity a corresponding general form, which is independent of the chance selection of the four variables. Corresponding to the ten functions $g_{\mu\nu}$, the gravitational potentials of the new theory, ten differential equations of the second order present themselves, without any additional hypothesis being necessary, and they have a corresponding form to that possessed by Poisson's equations for the Newtonian gravitational potential.

The latter allow the $g_{\mu\nu}$'s to be unambiguously calculated for a given gravitational field, i.e. when the distribution of the masses and the energy is given. The motion of every single body in the gravitational field under discussion then takes place according to the equation:

$$\delta \{\int ds\} = 0, \text{ where } ds^2 = \sum_1^4 g_{\mu\nu} dx_\mu dx_\nu.$$

This theory, which is built up from the most general assumptions, leads, for a first approximation, to Newton's laws of motion. Wherever deviations from the theory hitherto accepted reveal themselves, we have possibilities of testing the new theory experimentally. Before we turn to this question, let us look back, and become clear as to the attitude which the general theory of relativity compels us to adopt towards the various questions of principle we have touched upon in the course of this essay.

(*b*) RETROSPECT

1. The conceptions "inertial" and "gravitational" (heavy) mass no longer have the absolute meaning which was assigned to them in Newton's mechanics. The "mass" of a body depends, on the contrary, exclusively upon the presence and relative position of the remaining bodies in the universe. The equality of inertial and gravitational mass is put at the head of the theory as a rigorously valid principle. The hypothesis of equivalence hereby supplements the deduction of the special theory of relativity, that all energy possesses inertia, by investing all energy with a corresponding gravitation. It becomes possible—on the basis (be it said) of certain special assumptions into which we cannot enter here—to regard

rotations unrestrictedly, as relative motions too, so that the centrifugal field around a rotating body can be interpreted as a gravitational field, produced by the revolution of all the masses in the universe about the non-rotating body in question. In this manner mechanics becomes a perfectly general theory of relative motions.—As our statements are concerned only with observations of relative motions, the new mechanics fulfils the postulate that in physical laws observable things only are to be brought into causal connection with one another. It also fulfils the postulate of continuity; since the new fundamental laws of mechanics are differential laws, which only contain the line-element ds and no finite distances between bodies.

2. The principle of the constancy of the velocity of light *in vacuo*, which was of particular importance in the special theory of relativity, is no longer valid in general. It preserves its validity only in regions in which the gravitational potentials are constant, finite extents of which we can never meet with in reality. The gravitational field upon the earth's surface is certainly so far constant that the velocity of light, within the limits of accuracy of our measurements, had to appear to be a universal constant in the results of Michelson's experiments. In a gravitational field however, in which the gravitational potentials vary from place to place, the velocity of light is not constant; the geodetic lines, along which light propagates itself, will thus in general be curved. The proof of the curvature of a ray of light, which passes by in close proximity to the sun, offers us one of the most important possibilities of confirming the new theory.

3. The greatest change has been brought about by the general theory of relativity in our conceptions of space and time*.

According to Riemann the expression for the line-element, viz.

$$ds^2 = \sum_1^4 g_{\mu\nu}\, dx_\mu\, dx_\nu,$$

determines in our case the measure-relations of the continuous space-time manifold; and according to Einstein the coefficients $g_{\mu\nu}$ of the line-element ds have in the general theory of relativity

* This aspect of the problem has been treated with particular clearness and detail in the book *Raum und Zeit in der gegenwärtigen Physik* by Moritz Schlick, published by Jul. Springer, Berlin. The Clarendon Press is publishing an English rendering under the title: Space and Time in Contemporary Physics.

the significance of *gravitational potentials*. Quantities, which hitherto had only a purely geometrical import, for the first time became animated with physical meaning. It seems quite natural that gravitation should herein play the fundamental part, viz. that of predominating over the measure-laws of space and time. For there is no physical event in which it does not cooperate, inasmuch as it rules wherever matter and energy come into play. Moreover it is the only force, according to our present knowledge, which expresses itself quite independently of the physical and chemical constitution of bodies. It therefore without doubt occupies a unique position, in its outstanding importance for the construction of a physical picture of the world.

According to Einstein's theory, then, gravitation is the "inner ground of the metric relations of space and time" in Riemann's sense (*vide* the final paragraph of Riemann's essay "On the hypotheses which lie at the bases of geometry" quoted on page 15). If we uphold the view that the space-time manifold is continuously connected, its measure-relations are not then already contained in its definition as being a continuous manifold of the dimensions "four." These have on the contrary yet to be gathered from experience. And it is, according to Riemann, the task of the physicist finally to seek the inner ground of these measure-relations in "binding forces which act upon it." Einstein has discovered in his theory of gravitation a solution to this problem, which was presumably first put forward in such clear terms by Riemann. At the same time he gives an answer to the question of the true geometry of physical space, a question which has not remained silent for the last century,—but an answer, it is true, of a sort quite different from that which had been expected.

The alternative, Euclidean or non-Euclidean geometry, is not decided in favour of either one or the other; but rather space, as a physical thing with given geometrical properties, is banished out of physical laws altogether: just as ether was eliminated out of the laws of electrodynamics by the Lorentz-Einstein special theory of relativity. This, too, is a further step in the sense of the postulate that only observable things are to have a place in physical laws. The inner ground of metric relations of the space-time manifold, in which all physical events take place, lies, according to Einstein's view, in the gravitational conditions. Owing to the continual motion of bodies relatively to one another, these gravitational conditions are continually altering; and therefore one cannot speak of an invariable given geometry of measure or distance—whether Euclidean or non-Euclidean. As the laws of physics preserve their form in the general theory of relativity, independent of how the four variables $x_1, \dots x_4$ may chance to be chosen, the latter have no absolute physical meaning. Accordingly x_1, x_2, x_3, for instance, will not in general denote three distances in space which can be measured with a metre rule, nor will x_4 denote a moment of time which can be ascertained by means of a clock. The four variables have only the character of numbers, parameters, and do not immediately allow of an objective interpretation. Time and space have therefore not the meaning of real physical things in the description of the events of physical nature.

4. The gravitational theory, which follows out of the general theory of relativity, is, in contradistinction to the Newtonian theory, built up, not upon an elementary law of the *gravitational forces*, but upon an elementary law of the *motion* of a body in the *gravitational field*. Consequently, the expressions which would be *interpreted* as gravitational forces in the new theory play only a minor part in the building-up of the theory (as indeed the conception of force in mechanics altogether is to be regarded as only an auxiliary or derived conception, if we regard it as the object of mechanics to give a flawless *description* of the motions occurring in physical events).

Nor does Einstein's theory endeavour to explain the *nature* of gravitation; it does not seek to give a mechanical model, which would symbolise the gravitational effect of two masses upon one another. This is what the various theories involving ether-impulses attempted to do, by freely using hypothetical quantities which had never been actually observed, such as ether-atoms. It is very doubtful whether such endeavours will ever lead to a satisfactory theory of gravitation. For, the difficulties of Newton's mechanics are not contained only in the fact that it formulates the law of gravitation as a law of forces acting at a distance. Two much more serious points are:—first, that the close relationship existing between inertial and gravitational phenomena receives no recognition whatsoever, although Newton was already aware of the fact that inertial and gravitational mass are equal; and second, that Newton's mechanics does not present us with a theory of the relative motions of bodies, although we only observe relative motions of bodies with respect to one another. Re-moulding Newton's law of gravitational force, in order to make the attraction of matter more feasible, would therefore not have helped us finally to a satisfactory theory of the phenomena of motion (*vide* Note 25).

What distinguishes the Newtonian theory, above all, is the extraordinary simplicity of its mathematical form. Classical mechanics, which is built up on Newton's initial construction, will, for this reason, never lose its importance as an excellent mathematical theory for arithmetically following the observed phenomena of motion.

Einstein's theory, on the other hand, as far as the uniformity of its conceptual foundations is concerned, satisfies all the conditions for a physical theory. The fact that (by abandoning the Euclidean measure of distance) it cuts its connection with the familiar representation by means of Cartesian coordinates, will not be felt to be a disturbing factor, as soon as the analytical appliances, which have been called into use as a help, have been more generally adopted. It is impossible to say at present whether the new theory will considerably facilitate the practical problem of determining the path of a heavenly body.

The first task which falls to the lot of the astronomer is to test the theory experimentally in those phenomena in which measurable deviations from the results of the classical theory occur.

§ 6

THE VERIFICATION OF THE NEW THEORY BY ACTUAL EXPERIENCE

As far as can be seen at present, there are three possible experiments for verifying Einstein's theory of gravitation; all three can be performed only by the agency of astronomy. One of them —arising from the deviation of the motion of a material point in the gravitational field according to Einstein's theory, as compared with that required on Newton's theory—has already decided in favour of the new theory; the decisions of the other two, which are derived from the combination of electromagnetic with gravitational phenomena, will not be so readily forthcoming.

Since the sun far exceeds all other bodies of the solar system in mass, the motion of each particular planet is primarily conditioned by the gravitational field of the sun. Under its action the planet describes, according to Newton's theory, an ellipse (Kepler's law), the major axis of which—defined by connecting the point of its path nearest the sun (perihelion) with the farthest point (aphelion)—is at rest, relative to the stellar system. Upon this elliptic motion of a planet there are superimposed more or less considerable influences (disturbances) due to the remaining planets, which do not however appreciably alter the elliptic form; these influences partly only call forth periodical fluctuations in the defining elements of the initial ellipse (i.e. major axis, eccentricity, etc.), partly cause a continual increase or decrease of the latter. In this latter kind of disturbance are also to be classed the slow *rotation* of the major axis, and consequently also of the corresponding perihelion, relative to the stellar system; which has been observed in the case of all planets. For all the larger planets, the *observed* motions of the perihelion agree with those *calculated* from the disturbing effects (except for small deviations which have not been definitely established, as in the case of Mars); on the other hand, in the case of Mercury the calculations give a value which is too small by 43″ per hundred years. Hypotheses of the most diverse description have been evolved to explain this difference; but all of them are unsatisfactory. They oblige one to resort to still unknown masses in the solar system: and, as all the searches for masses large enough to explain this anomalous behaviour of Mercury prove fruitless, one is compelled to make assumptions about the distribution of these hypothetical masses, in order to excuse their *invisibility*. In view of these circumstances, there is no shade of probability in these hypotheses.

According to Einstein's theory, a planet, at the distance of Mercury for instance, moves, under the action of the sun's attraction, along the "straightest path," according to the equation:

$$\delta \{\int ds\} = \delta \left\{\int \sqrt{g_{11}dx_1^2 + g_{12}dx_1dx_2 + \dots + g_{44}dx_4^2}\right\} = 0.$$

The $g_{\mu\nu}$'s can be derived from the differential equations, which were given for them above, and which result from the *assumed sole presence of the sun and the planet being regarded as a mass concentrated at a point*. Einstein's developments give the ellipse of Kepler too as a first approximation for the path of the planet: at a higher degree of approximation, however, it transpires that the radius vector from the sun to the planet, between two consecutive passages through perihelion and aphelion, sweeps out an angle, which is about 0·05″ greater than 180°; so that, for each complete revolution of the planet in its path, the major axis of the path—i.e. the straight line connecting perihelion with aphelion—turns through about 0·1″ in the sense in which the path is described. Therefore, in 100 years—Mercury completes a revolution in 88 days—the major axis will have turned through 43″. The new theory, therefore, actually explains the hitherto inexplicable amount, 43 seconds per 100 years, in the motion of Mercury's perihelion, merely from the effect of the sun's gravitation. (The deviations due to such disturbances would only differ very inappreciably from the values obtained by Newton's theory in the case of the remaining planets.) The only arbitrary constant which enters into these calculations is the gravitational constant which figures in the differential equations for the gravitational potentials $g_{\mu\nu}$, as has already been mentioned on page 31. This achievement of the new theory can scarcely be estimated too highly.

The reason that a measurable deviation from the results according to Newton's theory occurs in the case of Mercury, the planet nearest to the sun, but not in the case of the planets more distant from the sun, is that this deviation decreases rapidly with increasing distance of the planet from the sun, so that it already becomes imperceptible at the distance of the earth. In the case of Venus, the eccentricity of the path is unfortunately so small, that it scarcely differs from a circle: and the position of the perihelion can therefore only be determined with great uncertainty.

Of the other two possibilities of verifying the theory, one arises from the influence of gravitation upon the time an event takes to pass. How such an influence can come about, will be evident from the following example: According to the new theory, an observer cannot immediately distinguish whether a change, which he observes during the passage of a certain event, is due to a gravitational field or to a corresponding acceleration of his place of observation (his system of reference). Let us assume an invariable gravitational field, denoted by parallel lines of force in the negative direction of the z-axis, and having a constant value γ for the acceleration with which all bodies in the field fall (i.e. characterised by conditions which approximately exist on the surface of the earth). According to Einstein's theory, any event will take place in this field in just the same way as it appears to occur when referred to a coordinate system which has an acceleration γ in the positive direction of the z-axis. Now if a ray of light, the time of oscillation of which is ν_1, travels from a point A—which is to be conveniently supposed at rest relatively to the corresponding coordinate system at the moment of departure of the ray—in the direction of the z-axis for a distance h to a point B: then an observer at B will, owing to his own acceleration, γ, have attained a velocity $\gamma \,.\, h/c$ at the instant the ray of light reaches him (c denotes the velocity of light). According to the usual Doppler Principle, he will assign a time of oscillation $\nu_2 = \nu_1 (1 + \gamma \,.\, h/c^2)$ to the ray of light as a first approximation, instead of ν_1. If we transfer the same event to the equivalent gravitational field, this result assumes the following form: The time of oscillation ν_2 of a ray of light at a place B, the gravitational potential of which differs from that of a place A by the amount ϕ, is connected with the time of oscillation there observed by the relation:

$$\nu_2 = \nu_1 (1 + \phi/c^2),$$

according to Einstein's theory of gravitation.

This special case shows how the duration of an event is to be understood as being dependent upon the gravitational condition.

Moreover, one can regard every vibrating system (which emits a spectral line) as a *clock*, the motion of which, according to the investigation made just above, depends upon the gravitational potentials of the place where it is stationed. This same "clock" will have a different time of oscillation at another place in the field according to the gravitational potential, i.e. it will go at a different rate. Consequently, a particular line in the spectrum of the light which comes from the *sun*, e.g. an Fe-line (iron), must appear to be shifted in comparison with the corresponding line as produced by a source of light on the *earth*; the gravitational potential at the surface of the sun has, corresponding to the latter's great mass, a different value from that at the surface of the earth, and a definite time of oscillation (colour) is characterised in the spectrum by a definite position (Fraunhofer line). It has not yet been possible to observe this effect, which amounts to about 0·008Å* for a wave-length of 400 $\mu\mu$ with certainty; but there are diverse points of attack for a treatment of this question in the case of the fixed stars too, and also signs of the existence of such a gravitational effect. The establishment of this effect beyond all doubt is an important task for stellar astronomy.

The third and *particularly important inference from Einstein's theory* is the dependence of the velocity of light upon the gravitational potential, and the resultant curvature (based upon Huygens' principle) of a ray of light in passing through a gravitational field. The theory asserts that a ray of light, coming e.g. from a fixed star, and which passes in close proximity to the sun, has a *curved path*. As a consequence of this *curvature*, the star must appear displaced from its true position in the heavens by an amount which attains the value 1·7″ at the edge of the sun's disc, and decreases in proportion to the distance from the centre of the sun. But since a ray of light which comes from a fixed star and passes by the sun can be caught only when the light of the sun, which overpowers all else, by its brilliancy, is intercepted before its entrance into our atmosphere, only the rare moments of a total eclipse come into account for this observation and for the solution of the problem.

* Å = Ångström unit = 10^{-8} cm.

The solar eclipse of 29th May, 1919, during which photographs were taken at two widely apart stations, for the purpose of this test, has been reported as fully confirming the general theory of relativity*.

The experimental verification of Einstein's theory of gravitation has thus not reached completion. But if, in spite of this, the theory can, even at this early stage, justly claim general attention, the reason is to be found in the unusual unity and logical structure of the ideas underlying it. In truth, it solves, at one stroke, all the riddles, concerning the motions of bodies, which have presented themselves since the time of Newton, as the result of the conventional view about the meaning of space and time in the physical description of natural phenomena.

* The results were made public at the meeting of the Royal Society on the 6th Nov., 1919. H. L. B.

APPENDIX

Note 1 (page 1). The term "special" principle of relativity is to signify the following postulate: All systems of reference which are in uniform rectilinear motion with regard to one another can be used for the description of physical events with equal justification. This means: if physical laws assume a particularly simple form when referred to any particular system of reference, they will preserve this form when they are transformed to any other coordinate system which is in uniform rectilinear motion relatively to the first system.

The *mechanics* of Galilei and Newton asserts the same postulate for the laws of motion. The relativity-principle of Galilei and Newton, therefore, does not differ from the special principle of relativity of Lorentz and Einstein, as regards its significance for *mechanical* events. But the latter extends its range of validity so as to include *electrodynamical* laws, taking due account of the constancy of the velocity of light *in vacuo*. The equations of transformation which connect coordinate systems in uniform rectilinear motion relatively to one another (i.e. connect equivalent systems with one another) are therefore different for the two principles. If a system x', y', z', t', moves with uniform motion parallel to the x-axis of the system x, y, z, t, with a velocity v, the equations of transformation are, according to the relativity-principle of the mechanics of Newton and Galilei:

$$x' = x - vt, \quad y' = y, \quad z' = z, \quad t' = t,$$

whereas, according to the "special" principle of relativity they are:

$$x' = \frac{x - vt}{\sqrt{1 - \frac{v^2}{c^2}}}, \quad y' = y, \quad z' = z, \quad t' = \frac{t - \frac{v}{c^2}x}{\sqrt{1 - \frac{v^2}{c^2}}}$$

(where c denotes the velocity of light *in vacuo*).

These so-called Lorentz-transformations of the special theory of relativity degenerate into those of the Newton-Galilei theory, if we put $v/c = 0$ in them, i.e. if the velocity of light be regarded as infinitely great in comparison with the velocity v of the systems relatively to one another. This important point will be discussed again later in § 3 (*b*), page 16.

For the rest, the points of view which have led us to abandon the Galilei-Newton principle of relativity, in favour of the special principle of relativity due to Lorentz and Einstein, have in recent years often been dilated upon at great length. References:

A. Einstein. "Zur Electrodynamik bewegter Medien," *Annalen der Physik*, 4 Folge, Bd. 17, S. 891.

M. Laue. "Das Relativitätsprincip," *Sammlung*, "*Die Wissenschaft*," Bd. 38, Braunschweig, 1913.

A. Petzoldt. "Das Relativitätsprincip der Physiker," *Zeitschrift für positivistische Philosophie*, 2. Jahrgang, Nr. 1.

Note 2 (page 12). The facts that *every* pair of points (point-pair) in space have the same magnitude-relation (viz. the same expression for the mutual *distance* between them) and that with the aid of this relation, every point-pair can be compared with every other, constitute the characteristic feature which distinguishes space from the remaining continuous manifolds which are known to us. We measure the mutual distance between two points on the floor of a room, and the mutual distance between two points which lie vertically above one another on the wall, with the same measuring-scale, which we thus apply in any direction at pleasure. This enables us to "compare" the mutual distance of a point-pair on the floor with the mutual distance of any other pair of points on the wall.

In the system of tones, on the contrary, quite different conditions prevail. The system of tones represents a manifold of two dimensions, if one distinguishes every tone from the remaining tones by its pitch and its intensity. It is, however, not possible to compare the "distance" between two tones of the *same pitch* but *different intensity* (analogous to the two points on the floor) with the "distance" between two tones of *different pitch* but *equal intensity* (analogous to the two points on the wall). The measure-conditions are thus quite different in this manifold.

In the system of colours, too, the measure-relations have their own peculiarity. The dimensions of the manifold of colours are the same as those of space, as each colour can be produced by mixing the *three* "primary" colours. But there is no relation between two arbitrary colours, which would correspond to the distance between two points in space. Only when a third colour is derived by mixing these two, does one obtain an equation between these three colours similar to that which connects three points in space lying in one straight line.

These examples, which are borrowed from Helmholtz's essays, serve to show that the measure-relations of a continuous manifold are not already given in its definition as a *continuous* manifold, nor by fixing its dimensions. A continuous manifold generally allows of various measure-relations. It is only experience which enables us to derive the measure-laws which are valid for each particular manifold. The fact, discovered by experience, that the dimensions of bodies are independent of their particular position and motion, led to the laws of Euclidean geometry where *congruence* is the deciding factor in comparing various portions of space. These questions have been exhaustively treated by Helmholtz in various essays. References:

Helmholtz. "Ueber die tatsächlichen Grundlagen der Geometrie," *Wiss. Abh.* 2, S. 10.

Helmholtz. "Ueber die Tatsachen, welche der Geometrie zugrunde liegen," *Wiss. Abh.* 2, S. 618.

Helmholtz. "Ueber den Ursprung und die Bedeutung der geometrischen Axiome," *Vorträge und Reden*, Bd. 2, S. 1.

Note 3 (page 12). The postulate that finite rigid bodies are to be capable of free motions, can be most strikingly illustrated in the realm of two-dimensions. Let us imagine a triangle to be drawn upon a *sphere*, and also upon a *plane*: the former being bounded by arcs of great circles and the latter by straight lines; one can then slide these triangles over their respective surfaces at will, and can make them coincide with other triangles, without thereby altering the lengths of the sides or the angles. Gauss has shown that this is possible because the *curvature* at every point of the sphere (or the plane, respectively) has exactly the same value. And yet the geometry of curves traced upon a *sphere* is different from that of curves traced upon a *plane*, for the reason that these two configurations cannot be deformed into one another without tearing (*vide* Note 24). But upon *both* of them planimetrical figures can be freely shifted about, and therefore theorems of congruence hold upon them. If, however, we were to define a curvilinear triangle upon an egg-shaped surface by the three shortest lines connecting three given points upon it, we should find that triangles could be constructed at different places on this surface, having the same lengths for the sides; but these sides would enclose angles different from those included by the corresponding sides of the initial triangle, and consequently such triangles would not be congruent, in spite of the fact that corresponding sides are equal. Figures upon an egg-shaped surface cannot therefore be made to slide over the surface without altering their dimensions: and in studying the geometrical conditions upon such a surface, we do not arrive at the usual theorems of congruence. Quite analogous arguments can be applied to three- and four-dimensional realms: but the latter cases *offer no corresponding pictures to the mind.* If we demand that bodies are to be freely movable in space without suffering a change of dimensions, the "*curvature*" of the space must be the same at every point. The conception of curvature, as applied to any manifold of more than two dimensions, allows of strict mathematical formulation; the term itself only hints at its analogous meaning, as compared with the conception of curvature of a surface. In three-dimensional space, too, various cases can be distinguished, similarly to plane- and spherical-geometry in two-dimensional space. Corresponding to the sphere, we have a non-Euclidean space with constant positive curvature; corresponding to the plane we have Euclidean space with curvature zero. In both these spaces bodies can be moved about without their dimensions altering; but Euclidean space is furthermore infinitely extended: whereas "spherical" space, though unbounded, like the surface of a sphere, is not infinitely extended. These questions are to be found extensively treated in a very attractive fashion in Helmholtz's familiar essay: "Ueber den Ursprung und die Bedeutung der geometrischen Axiome" (*Vorträge und Reden*, Bd. 2, S. 1).

Note 4 (page 12). The properties, which the analytical expression for the length of the line-element must have, may be understood from the following:

Let the numbers x_1, x_2 denote any point of any continuous two-dimensional manifold, e.g. a surface. Then, together with this point, a certain "domain" around the point is given, which includes points all of which lie in the plane.—D. Hilbert has strictly defined the conception of a multiply-extended magnitude (i.e. a manifold) upon the basis of the theory of aggregates in his *Grundlagen der Geometrie* (page 177). In this definition the conception of the "domain" encircling a point is made to give Riemann's postulate of the *continuous* connection existing between the elements of a manifold a strict form.

Setting out from the point x_1, x_2 we can continuously pass into its domain, and at any point, e.g. $x_1 + dx_1$, $x_2 + dx_2$, enquire as to the "distance" of this point from the starting-point. The function which *measures* this distance will depend upon the values of x_1, x_2, dx_1, dx_2, and for every *intermediate* point of the path which has conducted us from x_1, x_2 to the point $x_1 + dx_1$, $x_2 + dx_2$ will successively assume certain continually changing, and, as we may suppose, continually *increasing*, values. At the point x_1, x_2 itself it will assume the value zero, and for every other point of the domain its value must be positive. Moreover, we shall expect to find that, if for any intermediate point, denoted by $x_1 + d\xi_1$, $x_2 + d\xi_2$, $d\xi_1 = \frac{1}{2}dx_1$ and $d\xi_2 = \frac{1}{2}dx_2$, the required function which measures the distance of this point from x_1, x_2, will, at this point, have a value half that of its value for the point $x_1 + dx_1$, $x_2 + dx_2$. Under these assumptions, the function will be homogeneous and of the first degree in the dx's; its value will then appear multiplied by that factor in proportion to which the dx's were increased. In addition, it must itself vanish if all the dx's are zero; and if they all change their sign it must not alter its value, which always remains positive. It will immediately be evident that the function

$$ds = \sqrt{g_{11}dx_1^2 + g_{12}dx_1dx_2 + g_{22}dx_2^2}$$

fulfils all these requirements; but it is by no means the only function of this kind.

Note 5 (page 14). But the expression of the fourth degree for the line element would not permit of any geometrical interpretation of the formula, such as is possible with the expression

$$ds^2 = g_{11}dx_1^2 + g_{12}dx_1dx_2 + \ldots + g_{33}dx_3^2,$$

which latter may be regarded as a general case of Pythagoras' theorem.

Note 6 (page 15). By a "discrete" manifold we mean one in which no continuous transition of the single elements from one to another is possible, but each element to a certain extent represents an independent entity. The aggregate of all whole numbers, for instance, is a manifold of this type, or the aggregate of all *planets* in our solar system, etc., and many other examples may be found; and indeed all finite aggregates in the theory of aggregates are such discrete manifolds. "Measuring," in the case of discrete manifolds, is performed merely by "counting," and does not present any special difficulties; as all manifolds of this type are subject to the same principle of measurement. When Riemann then proceeds to say: "Either, therefore, the reality which underlies space must form a discrete manifold, or we must seek the ground of its metric relations outside it, in binding forces which act upon it," he only wishes to hint at a possibility, which is at present still remote, but which must, in principle, always be left open. In just the last few years a similar change of view has actually occurred in the case of another manifold which plays a very important part in physics, viz. "*energy*"; the meaning of the hint Riemann gives will become clearer if we consider this example.

Up till a few years ago, the energy, which a body emanates by radiation, was regarded as a continuously variable quantity: and it was therefore attempted to measure its amount at any particular moment, by means of a continuously varying sequence of numbers. The researches of Max Planck have, however, led to the view that this energy is emitted in "quanta," and that therefore the "measuring" of its amount is performed by counting the number of "quanta." The reality underlying radiant energy, according to this, is a discrete and *not* a continuous manifold. If we now suppose that the view were gradually to take root that, on the one hand, all measurements in space only have to do with distances between ether-atoms; and that, on the other hand, the distances of single ether-atoms from one another can only assume certain definite values, all distances in space would be obtained by "counting" these values, and we should have to regard space as a discrete manifold.

Note 7 (page 16). C. Neumann. *Ueber die Prinzipien der Galilei-Newtonschen Theorie*, Leipzig, 1870, S. 18.

Note 8 (page 16). H. Streintz. *Die physikalischen Grundlagen der Mechanik*, Leipzig, 1883.

Note 9 (page 17). A. Einstein. *Annalen der Physik*, 4 Folge, Bd. 17, S. 891.

Note 10 (page 18). Ph. Frank und H. Rothe. *Annalen der Physik*, 4 Folge, Bd. 34, S. 825.

The conditions fulfilled by the general equations of transformation, which connect two systems S and S' moving uniformly and rectilinearly with the velocity q relatively to one another, are:

1. The equations of transformation constitute a linear homogeneous group in the variable parameter q. This means that the successive application of two systems of equations of transformation, one of which connects the system S with S' and the second S' with S''—whereby S moves with a constant velocity q relatively to S', and S' moves with a constant velocity q' relatively to S''—leads in turn to a system of equations of transformation which has the same form as the two original equational systems; the parameter q'', which will occur in the new equations, will depend in a perfectly definite manner upon q' and q.

2. The contractions of the lengths depend only upon the value of the

parameter q.—One must of course, from the very outset, take into account the possibility of the length of a rod, as measured from the system at rest, turning out differently from its length as measured in the moving system. Condition 2 now requires that, if contractions (i.e. alterations of length due to these different ways of determining it) reveal themselves, they are to be dependent only upon the *magnitude* of the velocity of the two systems relatively to one another, and not upon the direction of their motion in space. This postulate thus endows space with the property of isotropy, and corresponds practically to the postulate of § 3 (a), viz. that every line-element can be compared in length with every other line-element, irrespective of the position or direction of either.

An essential point is that the constancy of the velocity of light is *not* required in either of these conditions 1 and 2. The *distinguishing* property of a *definite* velocity, that of preserving its value throughout *all* such systems which emerge out of one another as a result of such transformations, is, on the contrary, a strict corollary of these two general conditions; and the result of Michelson's experiment was only the determination of the *value* of this special velocity, which could naturally be gained only from direct experience.

Note 11 (page 19). Minkowski was the first to call particular attention to this deduction of the special principle of relativity.

Note 12 (page 21). The term "inertial system" was originally not associated with the system, which Neumann attached to the hypothetical body Alpha. Nowadays it is generally understood to signify a rectilinear system of coordinates, relatively to which a point-mass, which is only subject to its own inertia, moves uniformly in a straight line. Whereas C. Neumann only invented the body Alpha, as an absolutely hypothetical configuration, in order to be able to formulate the law of inertia, later researches, especially those of Lange, tended to show that, on the basis of rigorous kinematical considerations, a coordinate system could be derived, which would possess the properties of such an inertial system. However, as C. Neumann and J. Petzoldt have demonstrated, these developments contain faulty assumptions, and give the law of inertia no firmer basis than the body Alpha introduced by Neumann.

Such an inertial system is determined by the straight lines which connect three point-masses infinitely distant from one another (and thus unable to exert a mutual influence upon one another) and which are not subject to any other forces. This definition makes it evident why no inertial system will be discoverable in nature, and why, consequently, the law of inertia will never be able to be formulated so as to satisfy the physicist. References:

C. Neumann. *Ueber die Prinzipien der Galilei-Newtonschen Theorie*, Leipzig, 1870.

L. Lange. *Berichte der Kgl. Sächs. Ges. d. Wissenschaften. Math.-phil. Klasse*, 1885.

L. Lange. *Die Geschichte der Entwickelung des Bewegungsbegriffes*, Leipzig, 1886.

H. Seeliger. *Ber. der Bayr. Akademie*, 1906, Heft 1.

C. Neumann. *Ber. der Kgl. Sächs. Ges. d. Wiss. Math.-phys. Klasse*, 1910, Bd. 62, S. 69 and 383.

J. Petzoldt. *Ann. der Naturphilosophie*, Bd. 7.

Note 13 (page 22). E. Mach. *Die Mechanik in ihrer Entwickelung*, 4 Aufl. S. 244.

Note 14 (page 23). The new points of view as to the nature of inertia are based upon the study of the electromagnetic phenomena of radiation. The special theory of relativity, by stating the theorem of the inertia of energy, organically grafted these views on to the existing structure of theoretical physics. The dynamics of cavity-radiation, i.e. the dynamics of a space enclosed by walls without mass, and filled with electromagnetic radiation, taught us that a system of this kind opposes a resistance to every change of its motion, just like a heavy body in motion. The study of electrons (free electric charges) in a state of free motion, e.g. in a cathode-tube, taught us likewise that these exceedingly small particles behave like inert bodies; that their inertia is not, however, conditioned by the matter to which they might happen to be attached, but rather by the electromagnetic effects of the field, to which the moving electron is subject. This gave rise to the conception of the apparent (electromagnetic) mass of an electron. The special theory of relativity finally led to the conclusion that to *all* energy must be accorded the property of inertia.

Every body contains energy (e.g. a certain definite amount in the form of heat-radiation internally). The inertia, which the body reveals, is thus partly to be debited to the account of this contained energy. As this share of inertia is, according to the special theory of relativity, relative (i.e. represents a quantity which depends upon the choice of the system of reference), the whole amount of the inertial mass of the body has no absolute value, but only a relative one. This energy-content of radiant heat is distributed throughout the whole volume of each particular body; one can thus speak of the energy-content of unit volume. This enables us to derive the notion of density of energy. The density of the energy (i.e. amount per unit volume) is thus a quantity, the value of which is also dependent upon the system of reference. References:

M. Planck. *Ann. der Phys.* 4 Folge, Bd. 26.

M. Abraham. *Electromagnetische Energie der Strahlung*, 4 Aufl. 1908.

Note 15 (page 23). The determination of the inertial mass of a body by measuring its weight is rendered possible only by the experimental fact that all bodies fall with equal acceleration in the gravitational field at the earth's surface. If p and p' denote the pressures of two bodies upon the same support (i.e. their respective weights), and g denote the acceleration due to the earth's gravitational field at the point in question, then $p = m \,.\, g$ dynes and $p' = m' \,.\, g$ dynes, respectively, where m and m' are the factors of proportionality, and are called the *masses* of the two bodies, respectively. As g has the same value in both equations, we have

$$\frac{p'}{p} = \frac{m'}{m},$$

and we can accordingly measure the *masses* of two bodies at the same place, by determining their *weights*.

Although Galilei and Newton had already known that all bodies at the same place fall with the same velocity (if the resistance of the air be eliminated), this very remarkable fact has not received any recognition in the foundations of mechanics. Einstein's principle of equivalence is the first to assign to it the position to which it is, beyond doubt, entitled.

Note 16 (page 24). Arguing along the same lines B. and J. Friedländer have suggested an experiment to show the relativity of rotational motions, and accordingly the reversibility of centrifugal phenomena (*Absolute and Relative Motion*, Berlin, Leonhard Simion, 1896). On account of the smallness of the effect, the experiment cannot at present be performed successfully; but it is quite appropriate for making the physical content of this postulate more evident. The following remarks may be quoted:

"The torsion-balance is the most sensitive of all instruments. The largest rotating-masses, with which we can experiment, are probably the large fly-wheels in rolling-mills and other big factories. The centrifugal forces assert themselves as a pressure which tends from the axis of rotation. If, therefore, we set up a torsion-balance in somewhat close proximity to one of these large fly-wheels, in such a position that the point of suspension of the movable part of the torsion-balance (the needle) lies exactly, or as nearly as possible, in the continuation of the axis of the fly-wheel, the needle should endeavour to set itself parallel to the plane of the fly-wheel, if it is not originally so, and should register a corresponding displacement. For centrifugal force acts upon every portion of mass which does not lie exactly in the axis of rotation, in such a way as to tend to increase the distance of the mass from the axis. It is immediately apparent that the greatest possible displacement-effect is attained when the needle is parallel to the plane of the wheel."

This proposed experiment of B. and J. Friedländer is only a variation of the experiment which persuaded Newton to his view of the absolute character of rotation. Newton suspended a cylindrical vessel filled with water by a thread, and turned it about the axis defined by the thread till the thread became quite stiff. After the vessel and the contained liquid had completely come to rest, he allowed the thread to untwist itself again, whereby the vessel and the liquid started to rotate rapidly. He thereby made the following observations. Immediately after its release the *vessel alone* assumed a motion of rotation, since the friction (viscosity) of the water was not sufficient to transmit the rotation immediately to the water. So long as this state of affairs prevailed, the surface of the water remained a horizontal plane. But the more rapidly the water was carried along by the rotating walls of the vessel, the more definitely did the centrifugal forces assert themselves, and drive the water up the walls, so that finally its free surface assumed the form of a paraboloid of revolution. From these observations Newton concluded that the *rotation* of the walls of the vessel *relative* to the water does not call up forces in the latter. Only when the water itself shares in the rotation, do the centrifugal forces make their appearance. From this he came to his conclusion of the absolute character of rotations.

This experiment became a subject of frequent discussion later: and E. Mach long ago objected to Newton's deduction, and pointed out that it cannot be straightway affirmed that the rotation of the walls of the vessel relative to the water is entirely without effect upon the latter. He regards it as quite conceivable that, provided the mass of the vessel were large enough, e.g. if its walls were many kilometres thick, then the free surface of the water which is at rest in the rotating vessel would not remain plane. This objection is quite in keeping with the view entailed by the general theory of relativity. According to the latter, the centrifugal forces can also be regarded as gravitational forces, which the total sum of the masses rotating around the water exerts upon it. The gravitational effect of the walls of the vessel upon the enclosed liquid is, of course, vanishingly small compared with that of all the masses in the universe. It is only when the water is in rotation relatively to all these masses that perceptible centrifugal forces are to be expected. The experiment of B. and J. Friedländer was intended to refine the experiment performed by Newton, by using a sensitive torsion-balance susceptible to exceedingly small forces in place of the water, and by substituting a huge fly-wheel for the vessel which contained the water. But this arrangement too can lead to no positive result, as even the greatest fly-wheel at present available represents only a vanishingly small mass compared with the sum-total of masses in the universe.

Note 17 (page 24). We use the term "field of force" to denote a field in which the force in question varies continuously from place to place, and is given for each point in the field by the value of some function of the place. The centrifugal forces in the interior and on the outer surface of a rotating body are so distributed as to compose a field of this kind throughout the whole volume of the body, and there is nothing to hinder us from imagining this field to extend outwards beyond the outer-surface of the body, e.g. beyond the surface of the earth into its own atmosphere. We can thus briefly speak of the whole field as the centrifugal field of the earth; and, as the centrifugal field, according to the older views, is conditioned only by the inertia of bodies, and not by their gravitation, we can further speak of it as an inertial field, in contradistinction to the gravitational field, under the influence of which all bodies which are not suspended or supported fall to earth.

Accordingly the effects of various fields of force are superposed at the earth's surface: (1) the effect of the gravitational field, due to the gravitation of the particles of the earth's mass towards one another, and which is directed towards the centre of the earth; (2) the effect of the centrifugal field, which according to Einstein's view can be regarded as a gravitational field, and the direction of action of which is outwards and parallel to the plane of the meridian of latitude; finally (3) the effect of the gravitational field, due to the various heavenly bodies, foremost amongst them, the sun and the moon.

Note 18 (page 24). Eötvös has published the results of his measurements in the *Mathematische und Naturwissenschaftliche Berichte aus Ungarn*, Bd. 8, S. 64, 1891.

Whereas the earlier investigations of Newton and Bessel (*Astr. Nachr.* 10, S. 97, and *Abhandlungen von Bessel*, Bd. 3, S. 217), about the attractive effect of the earth upon various substances, are based upon observations with a pendulum, Eötvös worked with sensitive torsion-balances.

The force, in consequence of which bodies fall, is composed of two components: first the attractive force of the earth, which (except for deviations which may, for the present, be neglected) is directed towards the centre of the earth; and second the centrifugal force, which is directed outwards parallel to the meridians of latitude. If the attraction of the earth upon two bodies of equal mass but of different substance were different, the resultant of the attractive and centrifugal forces would point in a different direction for each body. Eötvös then states: "By calculation we find that if the attractive effect of the earth upon two bodies of equal mass, but composed of different substance, differed by a thousandth, the directions of the gravitational forces acting upon the two bodies respectively would make an angle of 0·356 second, i.e. about a third of a second with one another"; and later:

"I attached separate bodies of about 30 grms. weight to the end of a balance-beam about 25 to 30 cms. long, suspended by a thin platinum wire in my torsion-balance. After the beam had been placed in a position perpendicular to the meridian, I determined its position exactly by means of two mirrors, one fixed to it and another fastened to the case of the instrument. I then turned the instrument, together with the case, through 180°, so that the body which was originally at the east end of the beam, now arrived at the west end: I then determined the position of the beam again, relative to the instrument. If the resultant weights of the bodies attached to both sides pointed in different directions, a torsion of the suspending wire should ensue. But this did not occur in the cases in which a brass sphere was constantly attached to the one side, and glass, cork or crystal antimony was attached to the other; and yet a deviation of $\frac{1}{60000}$th of a second in the direction of the gravitational force would have produced a torsion of one minute, and this would have been observed accurately."

Eötvös thus attained a degree of accuracy, such as is approximately reached in weighing; and this was his aim: for his method of determining the mass of bodies by weighing is founded upon the axiom that the attraction exerted by the earth upon various bodies depends only upon their mass, and not upon the substance composing them. This axiom had therefore to be verified with the same degree of accuracy as is attained in weighing. If a difference of this kind in the gravitation of various bodies exists at all, it is, according to Eötvös, less than a twenty-millionth for brass, glass, antimonite, cork, and less than a hundred-thousandth for air.

Note 19 (page 26). *Vide* also A. Einstein, "Grundlagen der allgemeinen Relativitätstheorie," *Ann. d. Phys.* 4 Folge, Bd. 49, S. 769.

Note 20 (page 27). The equation $\delta \{\int ds\} = 0$ asserts that the variation in the length of path between two sufficiently near points of the path vanishes for the path actually traversed; i.e. the path actually chosen between two such points is the shortest of all possible ones. If we retain the view of classical mechanics for a moment, the following example will give us the sense of the principle clearly: In the case of the motion of a point-mass, free to move about in space, the straight line is always the shortest connecting line between two points in space: and the point-mass will move from the one point to the other along this straight line, provided no other disturbing influences come into play (Law of inertia). If the point-mass is constrained to move over any curved surface, it will pass from one point to another along a geodetic line to the surface, since the geodetic lines represent the shortest connecting lines between points on the surface.—In Einstein's theory, there is a principle quite corresponding, only much more general. Under the influence of *inertia and gravitation* every point-mass passes along the geodetic lines of the space-time-manifold. The fact of these lines not, in general, being straight lines, is due to the gravitational field, in a certain sense, putting the point-mass under a certain constraint, similar to that imposed upon the freedom of motion of the point-mass by a curved surface. A principle in every way corresponding had already been installed in mechanics as a fundamental principle for all motions by Heinrich Hertz.

Note 21 (page 28). *Vide* A. Einstein, *Ann. d. Phys.* 4 Folge, Bd. 35, S. 898.

Note 22 (page 29). The expression "acceleration-transformation" means that the equations giving the transformation from the variables x, y, z, t to the system of variables x_1, x_2, x_3, x_4, which is the basis of our discussion, can be regarded as giving the relations between two systems of reference which are moving with an *accelerated* motion relatively to one another. The nature of the state of motion of two systems of reference relatively to one another finds its expression in the analytical form of the equations of transformation of their coordinates.

Note 23 (page 31). Two things are to be undertaken in the following: (1) the fundamental equations of the new theory are to be written in an explicit form, and (2) the transition to Newton's fundamental equations is to be performed.

1. From the equation of variation $\delta\,\{\int ds\} = 0$ where

$$ds^2 = \sum_1^4 g_{\mu\nu}\, dx_\mu\, dx_\nu,$$

we have, after carrying out the operation of variation, the four total differential equations:

$$(1) \quad \frac{d^2 x_\sigma}{ds^2} = \sum_{\mu\nu} \Gamma^\sigma_{\mu\nu} \frac{dx_\mu}{ds} \frac{dx_\nu}{ds} \quad (\sigma = 1, 2, 3, 4).$$

These are the equations of motion of a material point in the gravitational field defined by the $g_{\mu\nu}$*'s.*

The symbol $\Gamma^\sigma_{\mu\nu}$ here denotes the expression

$$-\frac{1}{2}\sum_\alpha g^{\sigma\alpha}\left(\frac{\delta g_{\mu\alpha}}{\delta x_\nu} + \frac{\delta g_{\nu\alpha}}{\delta x_\mu} - \frac{\delta g_{\mu\nu}}{\delta x_\alpha}\right).$$

The symbol $g^{\sigma\alpha}$ denotes the minor of $g_{\sigma\alpha}$ in the determinant

$$\begin{Bmatrix} g_{11}, \ldots, \ldots, g_{14} \\ \ldots\ldots\ldots\ldots\ldots \\ g_{41}, \ldots, \ldots, g_{44} \end{Bmatrix}$$

divided by the determinant itself.

The ten differential equations for the "gravitational potentials" $g_{\mu\nu}$ are:

$$(2) \quad \sum_\alpha \frac{d\Gamma^\alpha_{\mu\nu}}{dx_\alpha} + \sum_{\alpha\beta} \Gamma^\alpha_{\mu\beta} \Gamma^\beta_{\nu\alpha} = \kappa\,(T_{\mu\nu} - \tfrac{1}{2} g_{\mu\nu} T).$$

The quantities $T_{\mu\nu}$ and T are expressions which are related in a simple manner to the components of the stress-energy-tensor (which plays the part of the quantity exciting the field in the new theory in place of the density of mass). κ is essentially equal to the gravitational constant in Newton's theory.

The differential equations (1) *and* (2) *are the fundamental equations of the new theory*. The derivation of these equations is carried out in detail in the tract by A. Einstein, *The Foundations of the general Principle of Relativity*, J. A. Barth, Leipzig, 1916.

2. In order to obtain a connection between these equations and Newton's theory, we must make several simplifying assumptions. We

shall first assume that the $g_{\mu\nu}$'s differ only by quantities, which are small compared with unity, from the values given by the scheme:

$$\left\{\begin{matrix} g_{11} & g_{12} & g_{13} & g_{14} \\ g_{21} & g_{22} & g_{23} & g_{24} \\ g_{31} & g_{32} & g_{33} & g_{34} \\ g_{41} & g_{42} & g_{43} & g_{44} \end{matrix}\right\} = \left\{\begin{matrix} -1 & 0 & 0 & 0 \\ 0 & -1 & 0 & 0 \\ 0 & 0 & -1 & 0 \\ 0 & 0 & 0 & +1 \end{matrix}\right\}$$

These values for the $g_{\mu\nu}$'s characterise the case of the special theory of relativity, i.e. the case of the condition free of gravitation. We shall also assume that, at infinite distances, the $g_{\mu\nu}$'s tend to, and do finally, assume the above values; that is, that matter does not extend into infinite space.

Secondly, we shall assume that the velocities of matter are small compared with the velocity of light, and can be regarded as small quantities of the first order. The quantities

$$\frac{dx_1}{ds}, \frac{dx_2}{ds}, \frac{dx_3}{ds},$$

will then be infinitely small quantities of the first order, and $\frac{dx_4}{ds}$ will equal 1, except for quantities of the second order. From the equations defining the $\Gamma^{\sigma}_{\mu\nu}$ it will then be seen that these quantities will be infinitely small, of the first order. If we neglect quantities of the second order, and finally assume that, for small velocities of matter, the changes of the gravitational field with respect to time are small (i.e. that the derivatives of the $g_{\mu\nu}$'s with respect to time may be neglected in comparison with the derivatives taken with regard to the space-coordinates) the system of equations (1) assumes the form:

$$(1\,a) \quad \frac{d^2 x_\tau}{dt^2} = -\frac{1}{2}\frac{\delta g_{44}}{\delta x_\tau} \quad (\tau = 1, 2, 3).$$

This would be the equation of motion as already given by Newton's mechanics, if $\frac{1}{2}g_{44}$ be taken as representing the ordinary gravitational potential. It still remains to be seen what the differential equation for g_{44} becomes in the new theory, under the simplifying assumptions we have chosen.

The stress-energy-tensor, which excites the field, degenerates, as a result of our quite special assumptions, into the density of mass ρ:

$$T = T_{44} = \rho.$$

In the differential equations (2) the second term on the left-hand side is the product of two magnitudes, which, according to the above arguments, are to be regarded as infinitely small quantities of the first order.

Thus the second term, being of the second order of small quantities, may be dismissed. The first term, on the other hand, if we omit the terms differentiated with respect to time, as above (i.e. if we regard the gravitational field as "stationary"), reduces to:

$$-\frac{1}{2}\left(\frac{\delta^2 g_{44}}{\delta x_1^2}+\frac{\delta^2 g_{44}}{\delta x_2^2}+\frac{\delta^2 g_{44}}{\delta x_3^2}\right)=-\frac{1}{2}\Delta g_{44} \text{ for } \mu=\nu=4.$$

The differential equation for g_{44} thus degenerates into Poisson's equation:

$$(2\,a)\quad \Delta g_{44}=\kappa\rho.$$

Thus, to a first approximation (i.e. if one regards the velocity of light as infinitely great, and this is a characteristic feature of the classical theory, as was explained in detail in § 3 (*b*): if certain simple assumptions are made about the behaviour of the $g_{\mu\nu}$'s at infinity; and if the time-changes of the gravitational field are neglected) the well-known equations of Newtonian mechanics emerge out of the differential equations of Einstein's theory, which were obtained from perfectly general beginnings.

Note 24 (page 32). The theory of surfaces, i.e. the study of geometry upon surfaces, makes it immediately apparent that the theorems, which have been established for any surface, also hold for any surface which can be generated by distorting the first *without tearing*. For if two surfaces have a point-to-point correspondence, such that the line-elements are equal *at corresponding points*, then corresponding finite arcs, angles, and areas, etc. will be equal. One thus arrives at the same planimetrical theorems for the two surfaces. Such surfaces are called "*deformable*" surfaces. The necessary and sufficient condition that surfaces be continuously deformable is that the expression for the line-element of the one surface:

$$ds^2=g_{11}dx_1^2+g_{12}dx_1dx_2+g_{22}dx_2^2$$

can be transformed into that for the other:

$$ds'^2=g_{11}'dx_1'^2+g_{12}'dx_1'dx_2'+g_{22}'dx_2'^2.$$

According to Gauss, it is *necessary* that both surfaces have equal *measures of curvature*. If the latter is constant over the whole surface, as e.g. in the case of a cylinder or a plane, all conditions for the deformability of the surfaces are fulfilled. In other cases, special equations offer a criterion as to whether surfaces, or portions of surfaces, are deformable into one another. The numerous subsidiary problems, which result out of these questions, are discussed at length in every book dealing with differential geometry (e.g. Bianchi-Lukat)*. This branch of training, which was hitherto of interest only to mathematicians, now assumes very considerable importance for the physicist too.

Note 25 (page 37). One must avoid being deceived into the belief that Newton's fundamental law is in any way to be regarded as an *explanation* of gravitation. The conception of attractive force is borrowed from our muscular sensations, and has therefore no meaning when applied to dead matter. C. Neumann, who took great pains to place Newton's mechanics on a solid basis, glosses upon this point himself in a drastic fashion, in the following narrative, which shows up the weaknesses of the former view:

"Let us suppose an explorer to narrate to us his experiences in yonder mysterious ocean. He had succeeded in gaining access to it, and a remarkable sight had greeted his eyes. In the middle of the sea he had observed two floating icebergs, a larger and a smaller one, at a considerable distance from one another. Out of the interior of the larger one, a voice had resounded, issuing the following command in a peremptory tone: 'Ten feet nearer!' The little iceberg had immediately carried out the order, approaching ten feet nearer the larger one. Again, the larger gave out the order: 'Six feet nearer!' The other had again immediately executed it. And in this manner order after order had echoed out: and the little iceberg had continually been in motion, eager to put every command immediately and implicitly into action.

"We should certainly consign such a report to the realm of fables. But let us not scoff too soon! The ideas, which appear so extraordinary to us in this case, are exactly the same as those which lie at the base of the most complete branch of natural science, and to which the most famous of physicists owes the glory attached to his name.

"For in cosmic space such commands are continually resounding, proceeding from each of the heavenly bodies,—from the sun, planets, moons and comets. Every single body in space hearkens to the orders which the other bodies give it, always striving to carry them out punctiliously. Our earth would dash through space in a straight line, if she were not controlled and guided by the voice of command, issuing from moment to moment, from the sun, in which the instructions of the remaining cosmic bodies are less audibly mingled.

* Forsyth's *Differential Geometry*. H. L. B.

"These commands are certainly given just as *silently* as they are obeyed; and Newton has denominated this play of interchange between commanding and obeying by another name. He talks quite briefly of a mutual attractive force, which exists between cosmic bodies. But the fact remains the same. For this mutual influence consists in one body dealing out orders, and the other obeying them."

INDEX

For EU product safety concerns, contact us at Calle de José Abascal, 56–1°,
28003 Madrid, Spain or eugpsr@cambridge.org.

www.ingramcontent.com/pod-product-compliance
Ingram Content Group UK Ltd.
Pitfield, Milton Keynes, MK11 3LW, UK
UKHW040227290926
15471UKWH00001B/36
* 9 7 8 1 1 0 7 6 0 1 3 7 6 *